BEESWAX ALCHEMY

How to make your own candles,
soap, balms, salves, and
home décor from the hive

Petra Ahnert

QUARRY

Brimming with creative inspiration, how-to projects, and useful information to enrich your everyday life, Quarto Knows is a favorite destination for those pursuing their interests and passions. Visit our site and dig deeper with our books into your area of interest: Quarto Creates, Quarto Cooks, Quarto Homes, Quarto Lives, Quarto Drives, Quarto Explores, Quarto Gifts, or Quarto Kids.

© 2015 Quarry Books

First published in 2015 by Quarry Books, an imprint of The Quarto Group, 100 Cummings Center, Suite 265-D, Beverly, MA 01915, USA. T (978) 282-9590 F (978) 283-2742 www.QuartoKnows.com

Quarry Books titles are also available at discount for retail, wholesale, promotional, and bulk purchase. For details, contact the Special Sales Manager by email at specialsales@quarto.com or by mail at The Quarto Group, Attn: Special Sales Manager, 100 Cummings Center, Suite 265-D, Beverly, MA 01915, USA.

10 9 8

ISBN: 978-1-59253-979-6

Digital edition published in 2015 eISBN: 978-1-62788-160-9

Library of Congress Cataloging-in-Publication Data available

Design: Burge Agency
Photography: Dan Bishop Photography
Printed in China

DEDICATION

This book is dedicated to the bees and the warriors who see to their protection; and to my parents, who instilled in me a love and curiosity of creatures large and small.

CONTENTS

PREFACE

I started working with beeswax in 2003 when my boyfriend and I got our first hives, but my love affair with craft started back when I was a child. I am an only child born to German immigrants. My mother lived an adventurous life before meeting my dad. She trained as a kindergarten teacher in Germany and spent eight years living in Peru, teaching in a German school. It was there where she was inspired to learn to make things by hand and from scratch. By the time I came along, she had become the quintessential housewife—an excellent cook, accomplished seamstress, and avid gardener. Later she studied weaving, pottery, and lapidary at the local community college. She had a mantra that still echoes in my head today: "You can make that!" She usually reminded me of this when I was eyeing something at the mall that I wanted to buy with my allowance and usually, under her tutelage, I succeeded in making something I liked even better.

My father was trained as a machinist in Germany and had the very precise mind this type of work required. He spent most of his life working as a machinist in Germany and the United States, and later as an engineer. At home, he was "Mr. fix-it" and "Mr. build-it." He always had some sort of project that kept him occupied. When he and my mom first got married, money was tight, so he built all their furniture—bed and night stands in the bedroom, couch and coffee table for the living room, and a kitchen table. Being somewhat of a traditionalist, he encouraged me to spend my free time with my mother, helping her with cooking, cleaning, and gardening, but he was doing far more exciting things in the garage and workshop. So I would often join him to help with his projects and learn about woodworking and mechanical things such as lawn mowers and cars.

I am definitely a product of these two characters! I remember taking an interests-and-career test given by the guidance counselor in high school. I answered each question as honestly as I could and then waited for what seemed like weeks for the test results. The results came in the form of a number with a supporting pie chart that mapped the numbers around the outside, and each number represented a career path. My result was 99 and I searched the pie chart for a long time before I finally found it—right smack dab in the middle of the circle. I don't remember exactly what they called that category, but I think it was "unclassified" or something equally obtuse. I was furious! It was NO help at all! In the years since, I have come to love my 99 score, realizing that it all boils down to balance and desire. Basically, I can do whatever I want to do, if I put my mind to it. But to be truly happy at what I am doing, I must find the balance between the analytical and the creative. It took me five years of college and sixteen years of work as an engineer to realize I needed to include creative outlets in my life as well.

When I met my boyfriend, Karl, he was looking for direction in his life. A short time later, he met a beekeeper's wife at the library. She needed help printing out something and after Karl helped her, he mentioned his fascination with bees, and asked her if her husband ever needed help. She chuckled and said that most people were leery of working with bees so he could usually use a hand. The next week Karl was helping him, and by the end of the season he had a starter hive. He was hooked! The next year we ramped up, buying ten more colonies, and learned that keeping the bees alive was not all that easy! We did have some super yummy honey, though, and some awesome light yellow beeswax.

Right about the time that Karl discovered beekeeping, he also stumbled on my stash of handmade soaps. These soaps were the product of a two-year soap-making binge. Years earlier, my friend Laurie had called me up and asked if I wanted to learn to make soap. At the time, I told her no—I had too many hobbies already and didn't need to put my time and money into yet another one, but she persisted and about a month later, I called her up and said, "Fine! You win. Show me how." I went shopping for supplies and we scheduled a day for soap making. Together we made two or three batches of soap that day. "OH MY GOSH!!!!" I told her, "This is the coolest thing ever!" I was hooked! After she left, I went out and bought more supplies and made enough soap to last an army a lifetime! I handed the bars out to family and friends and hoarded them away. When I couldn't justify making any more, I stopped.

Karl quickly discovered that he liked my handmade soaps more than commercially available soaps, so when we started selling our honey at local farmers' markets, we decided to sell my soaps as well. I fine-tuned my recipe and peddled them along with our artisan honeys. It was a great match. With a surplus of beeswax, I started making other items such as lip balms, solid lotions, and salves. Encouraged by the success of my products, I continued adding to my line with creams and lotions, perfumes, and candles. With each product addition, I acquired new ingredients and new knowledge. I also learned the value of experimentation. Following a recipe that comes out of a book or off the Internet can yield a good product, but what makes it good? Having a firm knowledge of the ingredients and why they are in the recipe is extremely important.

Now I have found that my business satisfies me on all levels—I use my analytical skills to create and execute recipes and I use my creative skills to develop the scent and look of my products. All in perfect harmony.

INTRODUCTION TO BEESWAX

Beeswax is the miracle of the beehive. The comb is built up from nothing and serves as a house, a nursery, and a food pantry. Over the millennia, bees have figured out that by building their combs into hexagons, the combs hold the most amount of honey and require the least amount of wax. The combs also serve as the perfect area for a bee to undergo its metamorphosis from egg to bee.

So what is beeswax? In the simplest terms, it is a wax produced by honey bees of the genus Apis. Beeswax consists of at least 284 different compounds, mainly a variety of long-chain alkanes, acids, esters, polyesters, and hydroxy esters, but the exact composition of beeswax varies with location. It has a specific gravity of about 0.95 and a melting point of over 140°F (60°C).

More specifically, it is a wax that is secreted from eight wax-producing glands on the worker bee's abdomen. The wax is secreted in thin sheets called scales. The scales, when first secreted, looks a bit like mica flakes. They are clear, colorless, tasteless, and very brittle. Beeswax is typically produced by the younger house bees when they are between twelve and twenty days old. As the bee grows older and begins to collect pollen and nectar, these glands start to atrophy, but their ability to produce beeswax doesn't disappear completely. When bees swarm they will rapidly produce wax comb, since they need to quickly create a place for the queen to lay eggs and somewhere to store food.

To form the beeswax into honeycomb, the bees will hang in strings and as wax is extruded from the glands of the wax-producing bees it is passed between the legs and mouths of the bees that form the chain, being chewed and molded into shape along the way. The bees will then use this wax to build the familiar hexagon-shaped honey cells. It is during this process that the wax starts to develop its color and opacity. Depending on what kind of nectar and pollen come into the hive and is consumed by the bees, microscopic bits of the pollen and nectar remain and get added to the wax. It takes about 1,100 scales to make one gram of wax.

Under the right conditions—meaning there is an adequate supply of food and the ambient temperature within the hive is between 91°F and 97°F (33°C and 36°C)—worker bees can produce beeswax on demand. They achieve the right temperature on cooler spring days by clustering around the wax-producing bees when they are building comb.

The production of beeswax in the hive is very costly, however. It takes about 8.4 pounds (3.8 kg) of honey to create 1 pound (425 g) of beeswax. This honey could be used to feed the nonforaging bees or it could be saved for times when nectar is in short supply. For this reason, beeswax is often chewed off in one spot and placed where it is needed. The reusing of old comb also contributes to the color, since it may have been used for brood rearing or honey storage and may contain cocoon remains, propolis, or pollen.

Most of the wax that is commercially available is made from what beekeepers call "cappings." When bees produce honey, the foraging bee collects the nectar and stores it in one of her two stomachs (one stomach is reserved for honey collection and the other for personal digestion). The nectar in the honey stomach mixes with enzymes and when the bee returns to the hive she places it into a waiting cell. As more cells are filled with nectar, bees fan their wings to create airflow through the

hive, which helps dry out the nectar. By lowering the moisture content of the nectar to less than 19 percent, the bees are ensuring that the honey will not spoil. Then the bees systematically work their way across frames and across honey boxes, capping off each cell to prevent additional moisture loss.

When beekeepers harvest the honey, they remove the frames with honey from the hive and bring them to the honey house for processing. Since all the honey cells have wax caps on them, just adding the frames to a honey extractor would yield no honey. So beekeepers first remove the wax cap using either a hot knife or some sort of flail. The wax cappings are added to a capping tank and the frames are placed into the extractor to spin out the honey.

What a beekeeper does with the wax cappings depends to a certain degree on how many hives he has. In most cases, heat is applied to the cappings, allowing the honey and wax to liquefy and separate into

two layers—honey on the bottom and wax on the top. After several more filterings to remove residual honey and miscellaneous bee parts, the wax looks pretty clean and is generally ready to go.

Beekeepers also melt down old honey and brood comb in order to install clean wax and do general maintenance on the frames. Over the years, brood comb will have raised multiple cycles of bees and the cocoon from the larvae stage will have turned the comb a dark brown. Also, potential pathogens may have been introduced either from the environment or from bees carrying the pathogen with them. These pathogens can decimate a hive rather quickly, which is why beekeepers often replace the old brood comb with clean wax. While wax from cappings and honey combs is fairly pure, the wax from brood combs contains a wide assortment of "stuff" which may include cocoons from both bees and wax moths, excrement from bee larvae, mites, pollen, propolis,

and bee parts. All this extra stuff is called "slum gum," and removing the slum gum from the wax is a more involved process. One method is to put the brood combs into burlap sacks and then add the bag to a hot water bath. The melted wax will flow through the burlap and the slum gum will stay in the bag. Beekeepers then press the burlap sacks to release the rest of the trapped wax from the slum gum. Once most of the wax is pressed out, the slum gum is discarded and the wax is molded into 30–50 lb (14–23 kg) blocks. The resulting wax is usually significantly darker than the cappings wax, ranging from light brown to almost black. If this wax were to be used for something such as candles, it would give off an unpleasant smell. A lot of beekeepers turn this wax in to bee supply stores for credit toward "clean" wax or wax that has already been turned into foundation for inclusion into new frames. The bee supply stores ship this dark wax to commercial wax processing operations that have specialized equipment with carbon filters that remove the color from the wax. This process is far better than how wax was filtered in the past, when it was bleached using noxious chemicals to remove the color. Most of the white wax available today is achieved naturally using carbon filters instead of chemicals.

One drawback to the heavily refined, highly filtered wax is that the aroma and charm of beeswax (as well as many of its unique health advantages) actually come from the natural "contaminants," including honey, propolis, and pollen. Bleaching or advanced refining of beeswax to remove its color and fragrance, yields a product that is a bit bland.

BRIEF HISTORY OF BEESWAX

The relationship between bees and humans dates to the hunter–gatherer days when, armed with nothing but a long stick and a lot of resolve, men would knock down hives from trees and run, returning to the scene to harvest the honey when it was deemed safe. Later, humans discovered that using smoke from a burning stick helped to subdue the bees, making the job a bit easier. The usefulness of wax could very well have been discovered then. Although some of this is conjecture on my part, there are cave paintings in Valencia Spain dating back about 8,000 years, that show two people collecting honey and honeycomb from a wild bee hive. They used baskets and gourds to transport the honeycomb, and a series of ropes to reach the hive.

We know more about the ancient Egyptians and their relationship with beeswax. They recognized the value of beeswax in mummification and used it for the embalming process. They also used the wax to seal the coffin and make it air tight, further preserving the body. The Egyptians preserved their writings on papyrus and on cave walls using beeswax, and these writings have remained unchanged for more than 2,000 years. They even recognized the importance of beeswax in health, as prescriptions dating back to 1550 BC called for beeswax in various formulations. Ancient jewelers and artisans utilized the lost wax casting technique, which involves sculpting an object in beeswax, coating the object with clay, and then hardening the clay with heat. The heat melted the wax, leaving a clay shell that was a perfect replica of the beeswax sculpture. Molten metal was then poured into the clay shell and allowed to harden before the clay was removed.

Egyptian priests also created the first voodoo dolls, using beeswax to create figures resembling their enemies before ritually destroying them. Egyptians also loved perfumes and were reputed to have made perfumed unguents, the precursors to today's solid perfumes. They incorporated beeswax, tallow, and various aromatic substances infused in oil, such as myrrh, henna, cinnamon, thyme, sage, anise, rose, and iris. The unguents weren't sold as perfumes, but rather for a multitude of medical uses.

The Chinese also recognized the importance of beeswax. About 2,000 years ago, one of China's most famous books on medicine, *The Shennong Book of Herbs*, praised beeswax for its beneficial influence on blood and energy systems and attributed beeswax with beauty enhancement and anti aging properties. Beeswax was also recognized as an important ingredient in wound treatment and dietary supplement.

Beeswax candles were already used by the ancient Egyptians, ancient Greeks, and in Rome and China. Beeswax candles have been used in European churches since the beginning of Christianity. The Roman Catholic Church only allowed beeswax candles to be used in the church. Although this law is still valid today, candles are no longer required to be 100 percent beeswax. By the eleventh century, however, churches were using huge amounts of candles. They were able to maintain the necessary amount of beeswax in part by having apiaries in every monastery and abbey.

In the days of Marco Polo, beeswax was abundant and was often used to pay tribute to kings. But despite its abundance, beeswax candles were only in the hands of the rich; the poor had to suffer with tallow candles.

HOW BEESWAX IS USED TODAY

Today, when technology seems to trump all else, it is encouraging to see that a product with such a rich history is just as vital in the twenty-first century as it was long ago.

Our world is changing and the future of the honey bee is a big unknown. Between climate changes, farming practices, genetic modifications, and pesticide use, the future looks bleak, but there is a grassroots movement to change things for the better. Young people are rediscovering rural life and are turning agricultural wastelands back into viable farms. These farms are rich in crop diversity and typically abide by more organic farming practices.

Although beeswax doesn't have nearly the applications it had years ago, it still has many great uses. The following are just a few examples of the diverse ways that beeswax is being used today.

CANDLES

Although the metal molds that have been used for ages are still being used to make beeswax candles, there are also modern silicone molds that make more intricate shapes easy.

ORNAMENTS

Adorning the house with decorative beeswax elements is a tradition that seems to be catching a second wind, from molded shapes to waxed fall leaves, with scent added or just the natural beeswax smell.

LIP BALM

Although there are other modern ingredients that have been used to firm up lip balm and form an occlusive layer, nothing is better than beeswax for keeping lips soft and moisturized.

COSMETICS

Beeswax plays a huge role in the making of beauty and personal care products! Things like moisturizer, lotion bars, and even homemade mascara all contain beeswax, which helps protect and moisturize the skin. Beeswax is even used in hair pomades and dreadlock wax.

MEDICINAL CREAMS AND SALVES

Beeswax has been used as a thickener and occlusive for millennia.

FOUNDATION FOR NEW HONEYCOMB IN HIVES

Beeswax is great for replacing existing hive body and honey super frames with clean wax, giving bees a head start.

WATERPROOF LEATHER

Combine equal parts beeswax, neatsfoot oil, and tallow. Dip a rag into the mixture, and rub it onto a pair of leather boots or gloves.

LUBRICANT FOR FLY-FISHING LINES

Rub a little beeswax on fishing line to keep it from tangling.

LUBRICANT FOR FURNITURE, DOORS, AND WINDOWS

Use beeswax to lubricate the hinges of doors and windows.

RUST PREVENTATIVE FOR TOOLS

To nullify the effects of oxidation caused by moist air, simply brush the tools with a solution of $1/3$ pound (142 g) of beeswax melted with one quart of turpentine. This can also be used on bronze as well.

WAX FOR BOWSTRINGS

Beeswax helps waterproof, lubricate, and slicken bowstrings, making them easier to play.

FRYING PANS AND IRONS

Beeswax protects pans against oxidation, helps in the seasoning process, and improves nonstick quality of pans and irons.

SOAP MAKING

Beeswax helps to make a nice, firm bar of soap.

BEARD AND MUSTACHE WAX

Beeswax is used in commercial mustache creams, especially those used to stiffen or control a man's beard or mustache.

GRAFTING WAX

Plant propagators apply grafting wax either by itself or in combination with grafting tape or string to prevent the surfaces from drying out, to keep the scion and rootstock surfaces pressed tightly together until they grow together, and to keep out water and air and the accompanying plant pathogens they may carry, including bacteria and fungi. A "cold" grafting wax can be made out of four parts resin; two parts beeswax, and one part tallow.

CRAYONS

Make your own crayons with beeswax, soap, and artist pigments. Mix equal parts grated soap and beeswax. Melt the beeswax and grated soap in a double boiler and stir until melted and the mixture is smooth. Color the mixture with dry artist's pigment.

SEALING ON JAMS AND JELLIES

Use just like paraffin to seal the tops of jams and jellies.

"LOST-WAX" CASTING

This is a method of metal casting where a molten metal is poured into a mold that has been created by means of a wax model. Once the mold is made, the wax model is melted and drained away.

WOOD FILLER

Fill small holes and cracks with plain or tinted beeswax.

GLASS ETCHING

A mixture of beeswax and resin act as a resist or mask when etching glass.

CANDY INGREDIENT

Beeswax gives gummy bears and jelly beans their rubbery and chewy texture.

CANELÉS DE BORDEAUX

These French bakery confections with a mahogany brown, crunchy outside and a creamy, gooey inside are molded in individual fluted cups, traditionally made of copper. It is a two to three day process to make them, and beeswax is used as a mold release and is part of the secret to the crunchy outside. Canelés are a confection that can be maddening in its elusiveness and many have given up without having reached perfection. Although I haven't given up yet, I am still a far cry from making what I would call the perfect Canelé! Stay tuned.

In this chapter
you'll learn
—Wax varieties
—Insider tips for
working with wax

CHAPTER 1
THE WAX

Most beekeepers have a love–hate relationship with beeswax. When they need it, it is usually not there, or at least not in the form needed, such as beeswax foundation. When they have it, it requires quite a bit of processing to make it useable, so it is often set aside to be handled "later." The production of wax also removes honey from the hive food stores in order to feed wax-producing bees, effectively reducing the amount of honey yield, so to maximize honey production, beekeepers try to keep as much comb in the hive as possible. Beekeepers traditionally process their wax either during honey extraction or during the winter and early spring months when other beekeeping maintenance is complete.

CANDLE WAX

For things such as candles, especially pillar candles, the residual honey in the wax causes the wax to burn unevenly, clogging the wick. The best way to get the last of the honey out of the wax is to allow it to clarify in a heated double boiler or wax tank. Admittedly, this task is easier to accomplish with the wax tank than a double boiler, since the wax must remain in a liquid state for quite a while until all the honey has settled to the bottom. I usually let mine settle for a couple days. The best way to determine if it is done is by checking the clarity of the wax. When it is first melted, it has murkiness to it. As the honey settles, the wax begins to clarify. When the wax is clear, filter it through a clean piece of felt cloth and pour into molds to divide the wax into useable portions. I usually do a variety of different sizes, from 1-ounce (28 g) ingots up to 10–20 pound (4.5–9 kg) blocks that store easily, but are still small enough to fit into my wax melter. This ensures that I have the right size for whatever I am making. The resulting wax is still yellow and retains the signature honeylike scent, although the filtering may have lightened up the wax slightly.

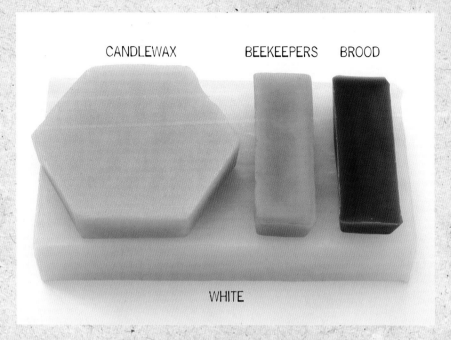

CANDLEWAX   BEEKEEPERS   BROOD

WHITE

BEEKEEPERS WAX

My first wax processing experience was while helping another beekeeper with his honey extracting. His operation seemed huge to me at the time (he maintains around 1,200 colonies), but I have learned that in comparison to other commercial operations, his is probably on the smaller side. There are some commercial beekeeping operations that maintain upwards of 50,000 colonies. There is very little difference between how these two types of beekeepers do their wax processing, except that the large-scale operations may have a more automated version of what I experienced.

The first step in the process is to place the honey frames into a machine that cuts the cap from the top of the honey cells. The removed cappings are funneled into a trough where they are pushed through a very coarse screen before entering the processing area. The frames are moved to the honey extractor. The wax processing area may have either a centrifuge where the honey is flung to the outside, rendering the cappings virtually free of honey, or a heated chamber with a series of baffles that separate the honey from the wax.

The wax that results from either of these processes is generally where the beekeepers stop. It is usually a really nice yellowish wax with a lovely honeylike smell. However, there is still some honey left in the wax. It may not be all that noticeable, but it is there. For some applications, such as lip balms, it is not a problem to a have a bit of honey in the wax, but for other applications, cleaner wax is needed.

BROOD WAX

This is the wax that comes from melting down old brood comb. Brood comb is usually replaced on a three- to five-year cycle. After the hive has reared multiple batches of brood on a frame of wax, the wax cells keep getting smaller and smaller, since the cocoon material remains. In the wild, Mother Nature has seen to the renewal of this comb, as the wax moth larvae loves to eat the brood comb. The moths are voracious eaters and can decimate a box of stored combs over the winter months, forcing the bees to build up the cell walls again before the queen can lay eggs into that space. For kept bees, beekeepers take on the role of the wax moth larvae by removing the old wax and replacing it with fresh wax foundation.

The process of rendering the wax from the cocoons, propolis, and other items in the brood comb is an arduous one. All these additional elements need to be filtered out before the wax itself can be cleaned. To do this, most beekeepers put the wax into burlap sacks, stacking press plates in between, and heat the whole works up to melt the wax. Once the wax is melted, the plates are pressed to release whatever wax remains in the slum gum and then the wax is strained off the top.

The wax that comes from this process is usually really dark. Although a lot of the bigger particles of propolis, cocoon, and slum gum are removed in this process, the microscopic bits still remain. Over time, these bits actually tint the wax, so even filtering the wax will only lighten it slightly. Although it may not look and smell as nice as yellow cappings wax, it does have its favored applications. It works really well for darker wood finishes. The light wax can have a "glazing" effect over time that is diminished when using the darker colored wax. It also works well on darker leathers. I use our dark wax to condition and polish my shoes. We also supply a local blacksmith with some dark wax. He prefers the darkest wax we have, since he uses it to seal and protect his wrought iron pieces.

WHITE WAX

Before I got into encaustic painting, I had no use for anything besides yellow wax. I loved the color, the scent, everything! So when I started making my first batch of encaustic medium, I used what I had on hand. I figured that I could make the yellow color part of my "signature" look. WRONG! Aside from my inept skills as an encaustic artist, it looked horrible. The colors were muddy and didn't flow as well as I had envisioned, so I did some research on "white" beeswax. My hope was that I would be able to remove some of the color from the wax myself. First I tried bleaching it in the sun. That works, but only to a certain degree. Sun does lighten the color of the wax, but the presence of microscopic bits of propolis, slum gum, and possibly honey added color that could not be completely bleached by the sun.

To remove the inherent colorants in the wax, it needs to undergo a different process that most at home cannot achieve. In the past, white wax was treated chemically to separate out the propolis and other "contaminants." This process yielded a nice, white wax, but traces of some of the chemicals remained. Not good! Today, most of the white wax is not treated chemically. It is run through a series of super fine filters that remove the "contaminants," without the addition of harsh chemicals.

The color of the resultant wax is very light and as close to white as possible, but there is also a translucency that is not available in the yellow waxes. This translucency is what makes it great to use for encaustics. Also, since a lot of the "contaminants" are what give the wax not only its signature color and smell, but also its myriad health benefits, white wax is not my preference for use on the skin.

INSIDER TIPS

RECIPE USAGE GUIDE

All of my recipes are written using percentages. The advantage to doing this is that the recipe can be easily scaled up or down. The disadvantage is that a little math is required to convert the equation into a usable recipe. Here's how to do that.

Example recipe:

Shea Butter	31%
Sweet Almond Oil	31%
Beeswax	31%
Flavor Oil	7%

1. Choose a unit of measure with which to work. Metric units, such as grams or milliliters, work best, as they are easier to multiply. They can be converted back to English units at the end of the process. The recipes in this book include gram measurements in addition to the percentages.

2. Decide the desired size of your batch. For this example we will use a 10-oz batch, which is equal to 283 grams.

3. Determine the percentage of each portion of the recipe.
Multiply the number of the full batch by the percentage for example, 283 g x 0.31(%) = 87.73 g.

4. Round to the desired precision (87.73 rounded to the nearest whole number = 88 g).

5. Finalize the recipe.
Using the sample recipe, it now looks like this:

Shea Butter	88 grams
Sweet Almond Oil	88 grams
Beeswax	88 grams
Flavor Oil	20 grams

6. Use a digital kitchen scale and a plastic deli container to measure ingredients. Remember to zero out the weight of the container before weighing ingredients.

SMALL AMOUNTS OF BEESWAX

When I need small amounts of beeswax for a recipe, it can be cumbersome to break a chunk of wax off a large block or grate some wax to use. Beeswax can be purchased commercially in the form of pastilles, which are small spherical beads of wax. Since most of this wax is the highly processed "white wax," I prefer to mold my own wax in one of two ways. The goal is to have wax that melts quickly (thin pieces work well) and is easy to measure (small).

One method is to pour the melted wax into a mold that is traditionally used to shape chocolate; the mold forms a thin-block with score lines that allow the chocolate—or in our case the beeswax—to be broken at the score lines. There are approximately six squares to an ounce (28 g), (they vary a bit in depth and actual break lines). Any combination of squares should work for almost any recipe. I use this method to save any leftover wax after pouring candles.

The second method calls for pouring the melted wax onto a cookie sheet (with sides) that is lined with freezer paper. This

will yield a sheet of wax that is variable in thickness, but can be broken into smaller chunks easily. I use this method when I have wax that I want to transform into flakes.

USE THE FREEZER

Beeswax gets brittle when it is frozen, making it easy to break off chunks as needed. I store my wax in blocks that weigh around 35 pounds (15.9 kg). Although great for storage, these heavy blocks are too big to fit into my wax tank. Since I use the most wax in winter, and weather gets cold around here, I use the great outdoors as my "freezer," but a regular freezer can be employed to do the same thing.

Once my block is frozen, I put it in a heavy-duty garbage bag (I use the black plastic contractor bags) and drop it on the pavement behind my house. Usually, it doesn't take more than a couple of drops to create some good-size chunks that are perfect for larger projects or to melt down and mold into smaller chunks.

The freezer is also great for helping to release candles that are stuck in metal molds. Just leave the stubborn candle in the freezer for a half hour or so, and when the mold is removed the candle should slip out fairly easily.

PROTECT SURFACES

When using beeswax for projects that call for wax to be applied, either by brushing it on, using another type of tool to apply it, or dipping it in wax, make sure that the work surface is protected. It is a lot easier to clean up a sheet of paper than it is to clean wax off the countertop.

CLEANING UP UTENSILS

The joy of a newly made candle can quickly become frustration when faced with the task of cleaning up. Beeswax does not clean up easily in the "traditional" way. Soap and water won't touch the wax unless it is hot, and even then it will only smear the wax around. I recommend acquiring a few containers that will be used exclusively for melting beeswax. That way, there is no cleanup needed and no leftover beeswax in tomorrow's dinner. If a beeswax container does need to be cleaned, the best way I have found is to turn it into a balm by

adding a bit of oil, which effectively drops the melting temperature. Then wash with hot water and some grease-dissolving dish soap. To clean a waxy bowl, wipe the bowl with oil (any oil will do) and heat the bowl with a heat gun (low heat) or in the microwave if it is safe to do so, and wipe with a paper towel. Repeat until the bowl is relatively clean and then wash with dish soap and water.

For products such as balms and creams, I like to use new plastic deli containers that can be tossed once the

product is made. They are quite heat proof, food grade, sterile, and come in a variety of sizes. They also make pouring product into smaller containers relatively easy, since the plastic is malleable enough to pinch into a pour spout. I realize that this not the most eco-conscious thing to do and I reserve their use for things that require sterility, such as lotions and creams.

In this chapter you'll
learn how to make
— Rolled candles
— Molded candles
— Hand-dipped candles

CHAPTER 2
CANDLE MAKING

A well-made beeswax candle is a beautiful thing!
However, making a well-made candle is not an
afternoon's fun. There is a lot of molding and burn testing
that needs to be done, before perfection is reached.
I hope the directions and hints found in this chapter will
make your road to perfection shorter and more fulfilling
than mine.

ROLLED CANDLES

Rolled candles are a great starting point for candle making, since the process requires very few tools and the candles can be made in a wide range of colors. They can be made as tapers, pillars, square, or even as small, thin candles such as birthday candles.

ROLLED TAPER CANDLES

MATERIALS

1 beeswax sheet, also called beeswax foundation, 8" x 16" (20.5 cm x 40.5 cm)

2 pieces (9", or 23cm each) of square braid wicking (2/0 is a good size to start, might be good to also have 3/0 and 1/0 on hand)

Yield: 2 taper candles

1. Gently fold the beeswax sheet in half, bending it back and forth until it breaks to create two 8" (20.5 cm) squares.

2. Lay a 9" (23 cm) piece of wick along the raw edge of the sheet, with one end of the wick flush with the side of the beeswax sheet and the other extended out.

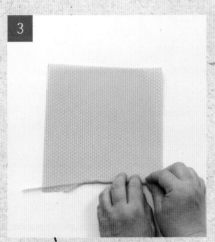

3. Pinch the beeswax over the wick, so that the wax is molded all around the wick from one end to the other.

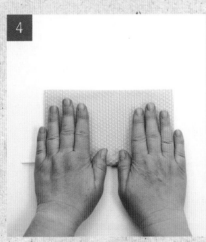

4. Start rolling the candle, keeping even pressure across the width. Don't crush the beautiful honeycomb pattern on the beeswax sheets, but make sure the candle doesn't come undone. If one side starts rolling tighter than the other, unroll the candle a bit and try again. It may take a couple tries to get it right.

5. Once the candle is evenly rolled, gently press the end edge into the candle to keep it in place.

ROLLED PILLAR CANDLES

Pillar candles can be made in a variety of shapes and sizes. Here is the basic information needed to expand on the rolled taper candle.

MATERIALS

1 full sheet beeswax

wicking (#2 will probably work, but also have #1 and #3 on hand)

scissors

Yield: 1 pillar candle

1. Gently fold the beeswax sheet in half lengthwise, bending it back and forth until it breaks to create two 4" x 16" (10 cm x 40.5 cm) pieces.

2. Cut the wick to approximately 5" (10.5cm) and follow directions for Rolled Taper Candles on page 23 to secure the wick and begin rolling the pillar.

3. When you reach the end of the first sheet of beeswax, lay the second sheet against the edge of the first sheet and continue rolling.

4. At the end of the sheet, gently press the end edge into the candle to keep it in place. The candle should measure approximately 2" (5 cm) across. Do a burn test (page 26) to make sure the correct wick size was used.

VARIATIONS

Although both taper and pillar candles use a rectangular or square piece of beeswax foundation, the beeswax can also be cut to create other shapes.

To make a candle with a tapered top, cut off a wedge piece from the top of the candle.

Rolling the trimmed off triangular piece of wax will result in a conical shape. For the holidays, I love to make a forest of small short stylized Christmas trees out of green wax using this technique.

A rolled candle can also be compressed into a square by flattening against your work surface then rotating the candle ninety degrees and flattening the next side. Continuing rotating and flattening until the candle has a nice square shape.

NOTE: This technique works better with pillar candles than tapers.

WICKING

Although there are lots of different kinds of candle wicking, for beeswax I like to use square braid cotton wicking. The wick acts as a pipeline that carries the melted wax in the form of a vapor to the flame via capillary action. Some wicks allow lots of fuel to flow quickly through a big pipe, while other wicks pump fuel more slowly through a smaller pipe. If you give the flame too much or too little fuel, it will either burn poorly or sputter out. The balance of fuel and flow needs to be just right.

The nomenclature of square braid wicking refers to the number of bundles, the ply of the wick, and how tightly it is braided. The 6/0 to 1/0 range of wicks, are constructed a bit differently than the larger wicks, but all of them are square, which helps to channel the wax fumes up to the flame. It is important to keep your wicks well-labeled and separated since similar sizes look identical. Often the only difference is the tightness of the braiding.

Square braid wick forms a carbon cap on the top of the wick. The carbon cap radiates heat outward from the flame, which helps melt wax that is further away from the flame. The wick also bends slightly as it burns, minimizing carbon build-up and making for a cleaner burn.

THE RIGHT SIZE WICK

Use the following test to determine the proper wick size and scale up or down as needed. Use this test should for all candles, although keep in mind that rolled candles can be a bit more forgiving. The test is especially important with molded pillar candles. Depending on the diameter of the candle, different wicks are needed to provide enough flame to melt the wax without creating a torch that threatens to burn down the house.

→

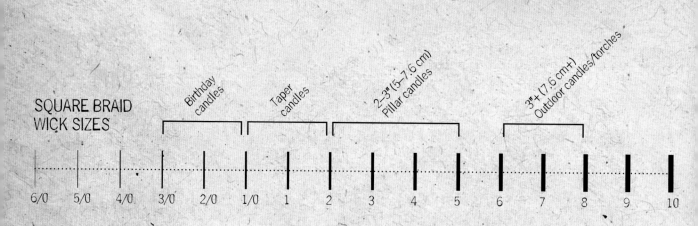

SQUARE BRAID WICK SIZES

Birthday candles | Taper candles | 2–3" (5–7.6 cm) Pillar candles | 3"+ (7.6 cm+) Outdoor candles/torches

6/0 5/0 4/0 3/0 2/0 1/0 1 2 3 4 5 6 7 8 9 10

BASIC BURN TEST

TEST THE PROPER WICK SIZE AND SCALE UP OR DOWN AS NEEDED FOR PILLAR OR VOTIVE CANDLES.

A. This wick is too small. (The flame is drowning in the small melt pool.)

B. Perfect

C. This wick is too large. (The flame is too large, resulting in a lot of smoke.)

1. Trim the wick to a length of ¼" (6 mm). If you are testing more than one wick, make sure the candles are clearly labeled.

2. Place the test candles on a clean, flat, heat-resistant surface about 3" to 6" (7.5 cm to 15 cm) apart. Be sure to select a draft-free spot that is full view of your workspace. Do not leave lit candles unattended.

3. Light the candles and record the time. It is critical to keep an eye on the candles while they are burning, especially when testing new wicks.

4. If testing pillar candles, allow them to burn for two hours then record the details of the melt pool and wick appearance. Ideally the melt pool will achieve the desired diameter by this point. Although there are individual preferences, I like my melt pool to extend almost out to the edge of the candle, leaving approximately ¾"–½" (6 mm–1.3 cm) of wax around the outside. If the flame has not melted enough of a melt pool, the wick is most likely too small. Note any soot or mushrooming on the wick.

5. Allow the candle to burn for another four hours and record the details of the melt pool and wick again before gently blowing out the flame. At this point the melt pool of a well-wicked candle will have achieved the desired diameter and should be approximately ½" (1.3 cm) deep. If the wick is mushrooming, the candle is sooting, or the melt pool is substantially deeper than ½" (1.3 cm), the wick is most likely too large.

6. Allow the candle to cool for at least five hours and repeat steps 3, 4, and 5 until the candle is completely burned. The quality of burn will almost always change during the entire burning of the candle. Burn the entire candle before deciding on a wick.

MOLDED CANDLES

All molded candles are basically made the same way and require the same basic tools, just different molds. It is important to use the cleanest wax possible when making molded candles. Impurities, such as residual honey, will affect the look and performance of the candle.

VOTIVE CANDLES

Votive candles and tealights are the workhorses of the candle world. Votives are intended to be burned in a votive holder, not on their own. This means that the candle should fit within the holder reasonably well and the candle should be wicked so that the wax is liquid all the way to the edge. Most votives burn for six to eight hours.

MATERIALS

1 lb (425 g) beeswax

Eco-2 pre-tabbed wicks

double boiler or wax melter

Kraft paper, newspaper, or butcher paper

silicone spray

votive molds

wick pins, enough for 10 candles

beeswax pouring pitcher

thermometer

heat gun

felt cloth, optional (enough to wrap around each of the molds once)

Yield: 8–10 votive candles

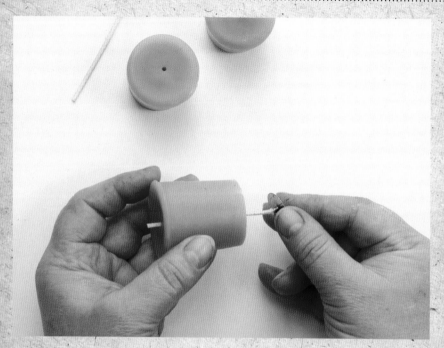

1. Melt the beeswax in a double boiler or a wax melter. Do not melt beeswax directly on the stove without the water bath. When I first started making candles, I used a clean coffee can to melt my wax. Periodically check the temperature of the wax to make sure it isn't getting too hot.

2. While waiting for the wax to melt, prepare the molds. If using metal molds, spray the inside with a mold release spray (silicone). Spray the wick pins as well and make sure they are properly seated on the bottom of the votive mold.

3. Once the wax is completely melted, use the thermometer to check the temperature of the wax. I like to pour my candles when the wax is in the 165°F–170°F (74°C–77°C) range. If it is hotter than that, let it cool a bit before pouring into the molds.

4. Warm the votive molds with the heat gun then pour the wax into the molds. Let the wax cool completely before trying to remove from the mold. Pull up on the wick pins to remove the candle and then tap the wick pin against a solid surface to dislodge it from the votive.

5. Thread a pre-tabbed wick up the hole left by the wick pin, trim the wick, and let it sit a day or two before doing a test burn.

MOLDED PILLAR CANDLES

Pillar candles are the longest burning candles. Even shorter 3" x 4" (7.5 cm x 10 cm) pillar candles have been known to burn for sixty hours or more, making them well worth the investment in beeswax.

MATERIALS

1 lb (425 g) beeswax

square braid wicking (try #3 for 2" (5 cm) diameter and #5 for 3" (7.5 cm) diameter molds)

pillar molds

silicone mold release spray

high-temperature metal taper (or other sealer such as wick putty)

beeswax pouring pitcher

thermometer

heat gun

felt cloth (enough to wrap around each of the molds once)

Yield: 2–3 small pillars or 1 large pillar

1. Melt the beeswax in a double boiler or a wax melter. Do not melt beeswax directly on the stove without the water bath. Check the temperature of the wax occasionally to make sure it isn't getting too hot.

2. While waiting for the wax to melt, prepare the molds. If using metal molds, spray the inside with a silicone mold release spray.

3. Most pillar candle molds come with a hole at the bottom of the mold to insert the wicking. I find it is easiest to dip a part of my wicking in beeswax and let it harden before trimming the dipped end at an angle to make it easier to thread the wick through the hole and far enough into the mold to allow me to retrieve it from the open end. A pair of pliers can help with this if the mold is a bit deeper than can be easily reached by hand.

4. Extend the wick about an inch (2.5 cm) or so above the candle mold. Use a wooden dowel, toothpick or a bobby pin to hold the wick in place and keep it centered. Wrap the wicking around the dowel and make sure it is tight inside the mold, but not stretching the wick. You'll also need to seal the spot where the wicking extends through the bottom of the mold. There are a number of products that can be used for this. My favorite is high-temperature metal tape, which is available in the heating section of home improvement stores.

TIP: I sometimes use another sealer called wick putty, a malleable dough that conforms to the bottom of the mold and seals the mold around the wick, preventing beeswax from leaking out.

5

5. Wrap the molds with a layer of felt cloth and configure them close together with taller candles in the center and shorter candles on the outside to make pouring the wax into the candle molds quick and efficient. I like to do a dry run with my pouring pitcher to make sure I don't have to pour the hot wax from a point that is too high up.

6. Once the molds are prepared, check on the wax. The target temperature should be in the range of 165°F–170°F (74°C–77°C). If it is hotter than that, let it cool a bit before pouring into the molds.

7. Before pouring the hot wax, make sure there is enough wax in the pouring pitcher to fill the mold completely. If the candle is poured in several stages, the line between pours will be visible.

Use a heat gun to quickly warm the inside of all the pillar molds and immediately pour the wax into the waiting molds. Fill the molds to about a ½" (1.3 cm) from the top.

7

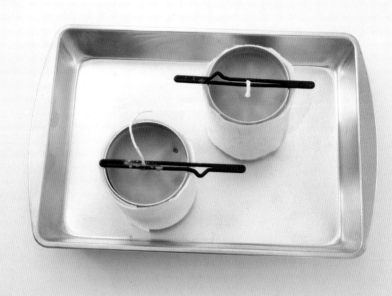

8

FILLING CAVITIES:
To counteract the cavity that forms as larger pillar candles cool, I make a relief hole or widen an existing hole with a nail or piece of toothpick, then I add some melted wax to the hole. It may not look like much of a hole, but you may be surprised by the amount of wax needed to fill the space. Repeat with more wax if necessary before it cools completely.

8. The wax will contract as it cools. How evenly the wax contracts depends on the ambient temperature of the room. Cooler temperatures will force the top of the candle to skin over and become a solid mass before the rest of the candle can contract sufficiently. The result is a cavity inside the candle. The bigger the wax mass, the longer it will take for the candle to cool and the greater the potential for a large cavity. This can be dangerous if left as is, since the wick can turn into a torch when it hits the open cavity.

9. Allow the candles to cool completely before trying to unmold. I like to do this the following day, especially if it is a larger diameter candle.

10. To unmold, first remove the tape or wick putty from the bottom of the mold. Then lightly tap the side of the mold against a soft surface, rotating the mold while tapping. Once the candle is loosened from the mold, tug gently on the wick and remove the candle. If it isn't cooperating, don't force it. Put the whole works in the freezer and let it cool for an hour or so. Then try again. This trick should work. Trim the wick flush at the bottom and to ¼ inch (6 mm) at the top.

11. The bottom of the candle may be bumpy and uneven. I use a dedicated electric skillet to melt and flatten candle bottoms. Start with the temperature of the skillet on low and see if the wax will melt. Electric skillets vary by manufacturer, so a bit of experimentation is necessary. If low doesn't melt the wax, increase the temperature until it starts to melt the wax relatively quickly. Place the candle in the skillet and spin it so that the high points in the candle bottom become more evident. Failure to do this will result in a leaning candle that will burn unevenly. Once the bottom is smoothed out and even, let it cool slightly and then clean up the edges with clean fingers.

12. Now the hard part: waiting for the candle to cure. Wait at least a day or two for the candles to cure before lighting them. This time frame allows the beeswax molecules to align and settle down. With pillar candles, a burn test is imperative to make sure that the right wick size is used. It sometimes takes five or more tries to get the wicking right, so don't get discouraged. Candles that are not burning correctly are not a waste. Simply melt them down the next time you are melting wax. The wick is not reusable.

HAND-DIPPED CANDLES

Hand-dipped candles have been around forever, probably because they are easy to make. A leisurely afternoon can yield a couple dozen passable tapers, especially if they don't have to be perfect. However, if the perfect taper is the goal, there is a bit more craft involved. I will begin with the basics and then delve into the details of creating beautiful candles.

HAND-DIPPED TAPER CANDLES

Hand-dipped taper candles are made from paper-thin sheets of wax that are added one layer at a time by dipping the wick repeatedly in hot wax. Most regular taper candles have twenty or more layers of wax.

MATERIALS

At least 5 lb (2.1 kg) beeswax (approximately 5 oz [142 g] per pair of tapers), including some small pieces and wax flakes

2/0 or 1/0 square braid wicking

tall melting pot with double boiler

scissors

nut or bolt to use as weight

paint sticks

THE IMPORTANCE OF THE CONTAINER

To make hand dipped candles, a tall wax melting container is important. The length of the finished candle is limited by the depth of the wax in the dipping container. A coffee can is great container to use, as it can be found around the house. The only disadvantage to using a coffee can is that it will probably only yield a 3–4" (7.5–10 cm)-long candle. To make taller candles, look for a taller, narrow metal container. An asparagus pot or something similar works well. There are really tall dipping pots available commercially, and I used one early on, but I had a problem keeping the wax hot from the bottom to the top. To maintain a constant wax temperature, a tall outer pot for the water bath is also required. It doesn't need to be as tall as the dipping pot, but it should be able to hold water to a height of at least two-thirds the level of the wax.

BEFORE YOU BEGIN

Make sure there is plenty of wax on hand. Remember that every time the wick is dipped in the wax, wax is removed from the pot. This means that once a few candles have been dipped, the wax level drops and candles will not be as tall as the first ones made. To remedy this problem, simply add more wax. I like to keep small pieces and flat wax flakes on hand for these occasions. Unlike big blocks of wax, they melt quickly and keep the process of making candles moving smoothly.

If you are dipping pairs, use the following general formula to cut the wicks into useable lengths:

(Height of desired finished candle + 2" [5 cm] loss on the bottom + 1" [2.5 cm] loss on top + ½" [1.3 cm] for wiggle room) x 2.

So to make 8" (20.5 cm) finished-length candles, cut the wick approximately 23" (58.5 cm) long.

In the above formula, I allowed a bit extra to tie a weight such as a washer or nut to the bottom of both wicks. This helps keep the wick straight through subsequent dippings, producing a beautiful, straight candle.

Temperature of the wax is very important. If the wax is too hot, the candles don't build up properly or build up more at the top than at the bottom. If the wax is too cool, the candles will look bumpy and have an exaggerated cone shape. It is best to play around with different dipping rhythms. How the candles look will depend on wax composition, temperature of the wax, temperature of the room, and individual style. Try dipping slowly on the way down and pulling it back out quickly. Then try the opposite. Also try leaving it at the bottom for an extra second or two. See how the shape of the candle changes with a couple of dips. Experimentation is the key.

1. Prepare a place to hang in-process candles while they cool. I like to use paint sticks, as they are wide enough to allow the candles to hang far enough apart to keep them from bumping into each other and the sticks are long enough to easily place them across a box or between two chair backs, allowing the candles to hang freely.

2. Prime all taper wicks by dropping a good quantity of wick into hot wax and letting it absorb completely before pulling it back out again. Watch for the bubbles and once they stop, wait a bit longer and then remove the wick. As it starts to cool, untangle the mass and start to straighten it out.

TIPS:
Save all the trimmings. The wax can be melted off and the weights reused.

Beeswax is very forgiving. If a candle doesn't turn out, put it back into the dipping vat. Once the wax has melted off the wicking, retrieve the wicking and start over.

3. Take the primed wick and use the formula above to cut to the desired length. Attach a nut or nail to the end of each end of the wick and bend the wick in half so that both sides are the same length.

4. Making sure that the wick is as straight as possible, hold the wick at the bend and dip the wick into the wax. There is no need to do this quickly. I dip the wick for about 3–4 seconds—1 second down, 1–2 seconds at the bottom, and 1 second up.

5. Before doing the subsequent dippings, I like to wait until the wax changes color slightly, approximately 3–4 minutes. If I am doing multiples, I hang up one and move on to the others and by the time I have completed one round with all the wicks, the first is ready for round two.

6. Continue dipping until the candles have reached about 75 percent of the desired thickness. Wax contracts as it cools, which means that a freshly dipped candle measuring ¾" (2 cm) will probably be thinner when it is completely cooled. Take this into account when making the candles. Wait for the wax to cool a bit longer and take a very sharp knife or scissors and cut the nut off the end of the wick.

7. Complete a couple more dips to finish the candle and hang up to cool completely. If the dipping vat is on the stove top, make the last dip a slightly hotter one (around 180ºF [82ºC]). This produces a smoother and shinier candle, enhancing the overall look of the finished taper.

HAND-DIPPED BIRTHDAY CANDLES

Birthday candles can be made in a couple different ways. Since they are simply really thin taper candles, they can be made in the same way, stopping when the approximate diameter reaches ½" (6 mm). But I prefer to make them look more like the traditional birthday candles—squared off at the top and bottom—so these candles are dipped first to a longer length and then trimmed to the desired length. Here is how I do it.

MATERIALS

beeswax

2/0 or 3/0 wicking

scissors

ruler

Yield: 10–15 longer candles (5–7"
[12.7–17.8 cm]) or 25–30 shorter
candles (3–4" [7.6–10.1 cm])

1. Cut and prime the wick, stretching it out as per tapers, page 38. There is no need to bother with a weight, since these candles will only be dipped 6–8 times total.

2. Dip the candles 5–6 times and measure the diameter. If it is close to the thickness desired, dip another time for good measure and let it cool until it just starts to turn from the pastel yellow to the more gold color, but is still a bit warm to the touch. With scissors, cut the wick just above and below the desired length of the candle.

3. To make the 7" (18 cm) candles shown above, cut candles into 7½" (19 cm) sections with a sharp pair of scissors.

4. Roll the cut sections on the table top between two sheets of waxed paper to straighten and smooth the imperfections. The candles also need to cool a bit more before continuing to the final trimming.

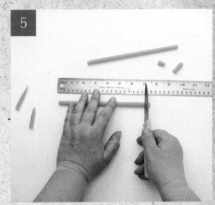

5. Measure 7" (18 cm) from one end of the candle and with a sharp knife, cut into the beeswax while rolling it to cut the wax, but not the wick. Pull the little chunk of wax off the wick and the candle is completed.

CHAPTER 3
BALMS AND BARS

Lip balm is one of the first things I made with our own beeswax and for good reason. Beeswax can help protect the lips from the harmful rays of the sun, has a pleasant smell, and it can help prevent infections and cold sores. In a pinch it can also be used to moisturize other body parts such as rough elbows or ragged cuticles.

BASIC LIP BALM

A lip balm in a tube needs to be stiff enough to hold its shape and should not melt if kept in a pants pocket, yet when applied to lips, it should glide on easily. Pour that same lip balm into a tin, and it will probably be too stiff to be picked up with a finger and applied to the lips. I prefer lip balm in tubes, and the following recipe is one of my favorites.

MATERIALS

yellow beeswax	27 g	30.00%
virgin coconut oil	26 g	29.60%
sweet almond oil	26 g	29.70%
castor oil	1 g	1.00%
lanolin	8 g	9.30%
Vitamin E oil	0.1 g	0.10%
rosemary oleoresin extract (ROE)	0.1 g	0.10%
lip-safe flavor oil	0.2 g	0.20%
wooden stir stick or kabob skewer		
knife		
digital scale		
double boiler		
20 lip balm tubes		
Yield: 20 tubes		

1. Coarsely chop the beeswax or use beeswax pastilles. Weigh the ingredients. Place the beeswax, oils, and lanolin into a small pot or heat-safe bowl. Gently heat in the top of a double boiler until the beeswax and oils have melted.

2. Once melted, remove from the double boiler and add essential oils and Vitamin E oil, stirring with a stir stick or wooden kebab skewer to combine. While it is still hot, pour the mixture into lip balm tubes. Allow to cool completely before placing caps onto the lip balm containers.

LIP BALM WITH BUTTERS

Here is a lovely variation that includes cocoa butter. Cocoa butter comes in either the "natural" scent, which some describe as somewhat chocolatelike, or deodorized. There is no difference between the two, except personal preference.

MATERIALS

yellow beeswax	20 g	21.64%
cocoa butter	20 g	21.64%
sunflower oil	45 g	48.70%
lanolin	5 g	5.41%
Vitamin E oil	2 g	2.16%
lip-safe flavor oil	12 drops	0.44%
knife		
digital scale		
double boiler		
20 lip balm tubes		
Yield: 20 tubes		

1. Coarsely chop the beeswax or use beeswax pastilles. Weigh ingredients. Place the beeswax, butter, sunflower oil, and lanolin in a small pot or heat-safe bowl. Gently heat the mixture in the top of a double boiler until the beeswax and butter have melted.

2. Once melted, remove from the stovetop and add essential oil and Vitamin E oil. Immediately pour the mixture into lip balm containers.

3. Allow to cool completely before placing caps onto the lip balm containers.

LIP GLOSS

Lip gloss is generally a softer balm that is stored in a jar or tin rather than a tube and usually has ingredients that add a bit of shine as well.

MATERIALS

yellow beeswax	10 g	9.87%
castor oil	20 g	19.74%
sweet almond oil	30 g	29.62%
shea butter	25 g	24.68%
cocoa butter	15 g	14.81%
lip safe flavor oil	0.3 g	0.30%
knife		
digital scale		
double boiler		
lip gloss jars or tins		

Yield: Makes twelve ¼ oz (7 g) jars, or six ½ oz (15 g) tins

1. Coarsely chop the beeswax or use beeswax pastilles. Weigh the ingredients. Place the beeswax, butter, and oils and put into a small pot or heat-safe bowl and gently heat in the top of a double boiler until the beeswax and butters have melted.

2. Once melted, remove from the stovetop and add essential oil. Immediately pour the mixture into lip gloss containers. Allow to cool completely before placing caps onto the lip gloss containers.

SOLID BODY-CARE BARS

Much like lip balms, solid body-care bars are very simple to make, but it can take a while to achieve just the right combination of butters, oils, and wax for both personal preference and regional weather conditions—a person in the tropics, for example, may want a completely different lotion than someone in the Arctic. Experimentation is key. My aim when making a solid lotion bar is to find a balance between emolliency and skin feel. Let's face it, solid lotion bars will leave skin feeling greasier than a liquid lotion, since there is no water to cut the greasy feeling, but with a good recipe, that greasiness can be significantly reduced.

SOLID LOTION BAR

(Image appears on page 47)

Solid lotion bars are an anhydrous blend of butters, oils, and waxes, not emulsified products that contain a large percentage of water.

This blend yields a bar that can be handled easily without making a mess, but melts on contact to soothe and create an emollient barrier. Although butters such as shea butter and cocoa butter are wonderful for the skin, they do not directly moisturize the skin. The moisture needs to come from other sources, such as the dampness that remains after hand washing. When these butters are paired with the beeswax, however, they work together to create an occlusive barrier that seals in that moisture. Various butters also contain fatty acids that help to nourish the skin.

MATERIALS

yellow beeswax	27 g	30.20%
virgin coconut oil	6 g	6.90%
sweet almond oil	12 g	13.80%
jojoba oil	12 g	13.80%
shea butter	10 g	10.50%
cocoa butter	12 g	10.50%
mango butter	5 g	5.30%
Vitamin E oil	0.1 g	0.10%
rosemary oleoresin extract (ROE)	0.1 g	0.10%
scent	0.9 g	1.00%
double boiler		
digital scale		
wooden stir stick		

clear plastic guest soap molds. (They hold about 1 oz (28 g) of balm each and the cavities are small enough that the resulting balms fit easily into a tin. Muffin tins can also be used by filling 1" (2.5 cm) deep.)

2–3 oz (57–85 g) shallow tins (optional, but a nice touch)

Yield: Makes approximately three 1 oz (28 g) lotion bars

1. Melt beeswax, coconut oil, sweet almond oil, and jojoba together in a double boiler on low to medium heat until the beeswax is completely liquefied. Add cocoa butter and mango butter and melt until liquefied and then remove from heat.

2. Slowly add small pieces of shea butter to the mixture and allow to melt.

3. After the shea butter has melted into the mixture, add the Vitamin E oil, ROE, and scent and pour into molds.

4. Put the molds with the lotion into the freezer for about 15 minutes and then invert the mold onto a clean work surface. The solid lotion bars should come out easily with a light tap.

5. Place one solid lotion bar in each of the tins.

MASSAGE BAR

MATERIALS

yellow beeswax	56 g	25%
cocoa butter	79 g	35%
sesame oil	88 g	39%
scent	2 g	1%
double boiler		
digital scale		
molds (clear plastic soap molds or muffin tins work well)		
tins with covers, large enough to hold each bar		
Yield: Makes four 2 oz (57 g) bars		

This massage bar is essentially a solid lotion made with the oils and butters that work well for massage.

With massage, it is more important to have enough slip and adequate working time rather than a product that doesn't feel greasy and absorbs quickly. I like to use a butter that creates both a hard bar and a bar that melts around body temperature, so cocoa butter is a stellar choice. Mold these in nice large pucks so that they are easy to use and can cover large areas.

1. Heat the beeswax with the sesame oil in a double boiler over low heat until melted, then add the cocoa butter.

2. Once the cocoa butter is melted, add scent, pour into molds, and chill for 15 minutes in the freezer.

3. Unmold the bars onto a clean work surface and place each bar in a tin or some other container to keep the bar clean.

CUTICLE BALM

My cuticles take a beating, so I like to keep cuticle balm stashed throughout the house and car, so it is available to me whenever I remember to apply it. This balm will help to keep cuticles looking nice while improving nails at the same time. I love to include lanolin and lecithin in my cuticle balm. These two ingredients act as barriers, preventing nails from absorbing too much water, and the phospholipids in the lecithin keeps the nail flexible and less prone to breakage.

1. Heat the beeswax, mango butter, and soybean oil together in a double boiler until completely melted.

2. Add the lanolin, lecithin, and Vitamin E oil, and when completely melted and mixed, add the essential oils.

3. This will be a stiff mixture when cooled, so I like to pour it into lip balm tubes. If more of a balmlike consistency is desired, cut back the beeswax to 15 percent and add the difference to the oils.

MATERIALS

yellow beeswax	29 g	28%
mango butter	32 g	30%
soybean oil	21 g	20%
lanolin	10.5 g	10%
lecithin	10.5 g	10%
Vitamin E oil	1 g	1%
lemon 5-fold essential oil	0.3 g	.33%
pink grapefruit essential oil	0.4 g	.34%
peppermint essential oil	0.3 g	.33%
double boiler		
digital scale		
spoon		
20 lip balm tubes		

Yield: Twenty 1.5 oz (42 g) tubes

HEEL BALM

My heels are especially prone to cracking, and applying this heel balm consistently over a two-week period always does the trick. This recipe has emu oil in it, which is supposed to penetrate skin and help to bring other ingredients into the skin as well.

MATERIALS

yellow beeswax	29 g	20%
shea butter	51 g	35%
coconut oil	36 g	25%
emu oil	12 g	8%
sweet almond oil	14.5 g	10%
Vitamin E oil	1.5 g	1%
peppermint essential oil	0.7 g	.5%
lemon 5-fold essential oil	0.7 g	.5%
double boiler		
digital scale		
wooden stir stick		
deli cup		
large twist-up tubes (I like oval mini deodorant tubes, but ½ oz [14 g] lip balm tubes work as well)		

Yield: Five 1 oz (28 g) tubes

1. Heat the beeswax, shea butter, and coconut oil together in a double boiler until completely melted. Remove from heat.

2. Add the emu, sweet almond, and Vitamin E Oil, stirring with a stir stick to ensure it is thoroughly mixed and all the oils are completely melted. Let the mixture cool slightly and add the essential oils. Stir to incorporate.

3. Transfer a small portion of the balm to the deli cup and return the rest to the double boiler to stay warm. Pinch the deli cup to form a pour spout and pour the balm into a tube. repeat until remaining tubes are filled. Allow to cool undisturbed.

4. When the balm has cooled, add the caps.

CHEST RUB

Although chest rubs can be purchased commercially, most are made with petrolatum. My version is a soft balm recipe utilizing essential oils that improve respiratory congestion. For very young children, use half the amount of essential oils.

MATERIALS

yellow beeswax	26 g	23%
olive oil	85 g	75%
Vitamin E oil	1 g	1%
eucalyptus essential oil	0.6 g	.5%
camphor essential oil	0.5 g	.35%
birch essential oil	0.1 g	.15%
double boiler		
digital scale		
spoon		
4 salve containers, 1-oz (28 g) each		
Yield: Makes approximately four 1 oz (28 g) portions		

1. Heat the beeswax and olive oil in a double boiler until melted.

2. Remove from heat and stir in the Vitamin E oil and essential oils.

3. Pour into salve containers.

NONPETROLEUM JELLY

Nonpetroleum Jelly is a product that looks and feels a lot like petroleum jelly, but it is made with nonpetroleum products. Castor oil and beeswax have a symbiotic relationship. When heated beeswax is whisked into heated castor oil and whipped to a jelly consistency, the mixture provides a protective barrier to the skin, holding in moisture and helping to prevent chapping or irritation.

MATERIALS

yellow beeswax	9 g	9%
castor oil	90 g	90%
Vitamin E oil	1 g	1%
double boiler		
digital scale		
pot		

Yield: Makes 100 grams, or approximately two 2 oz (57 g) jars

1. Melt the beeswax in a double boiler and warm the castor oil in a separate pot.

2. Add the castor oil and Vitamin E oil to the melted wax, gently whipping the mixture as it cools. When finished it should have a jellylike consistency.

3. Package jelly into jars.

CHAPTER 4
SALVES, CREAMS, AND SCRUBS

Salves consist of infused oils, beeswax, and at times, essential oils. By using herbs in different combinations, you can make salves that serve a variety of purposes. Making salves from scratch is not hard, but it can take some time. Keep in mind that healing herbs are delicate, and require a gentle hand. I like to take the slow method of room temperature infusion to extract the "good stuff" from the herbs. I feel this is the kindest and best way to extract the healing essences. Some people will use heat to speed up the process, and with a watchful eye, this can be done successfully, but it is very easy to scorch the herbs and negate any healing properties.

SALVE MASTER RECIPE

Since the only thing that differentiates one salve from another is the herb used, I use this a master recipe as a base, and I've provided a couple of favorite recipes that utilize that base recipe.

MATERIALS

yellow beeswax	40 g	20%
liquid oil, infused with herb	158 g	78.5%
Vitamin E oil	2 g	1%
essential oil blend	1 g	.5%
double boiler		
digital scale		
spoon or stir stick		
4 tins, 2 oz (56 g) each		

Yield: Makes enough to fill approximately four 2 oz (56 g) tins

1. Heat the beeswax and infused oil in a double boiler until the beeswax is completely melted. Take care not to get the oil too hot, so that there is no damage to the oil.

2. Remove from heat, and add Vitamin E oil and essential oils, if desired.

3. Stir to mix well and pour into waiting containers. If using plastic jars, make sure the mixture isn't too hot before pouring into jars, as they can deform with too much heat.

OIL HERBAL INFUSION

I like to use the "simplers" method to infuse my oils, meaning that the recipe is based on ratios, and measurements are referred to as "parts." This simple way of measurement allows the formulator to make the recipe in any volume desired.

MATERIALS

herbs of your choice

high-quality extra virgin olive oil

Vitamin E oil

canning jar with lid

BEFORE YOU BEGIN

For salves, use a high-quality oil, such as extra virgin olive oil. It is wonderful for skin all on its own and it is made even better with the addition of herbs. Other oils to try are grape seed oil, sweet almond oil, and apricot seed oil. For each of my recipes, I will suggest an oil. While it is not necessary to use that oil, understand that the oil was chosen for what it brings to the table.

Many herbs can be used in oil infusions. The combinations are limited only by the imagination. I recommend, however, that only dried herbs are used. It is especially important to use herbs that are completely dry, as any moisture within the oil will spoil the batch. If the materials are not completely dry, bacteria will flourish, resulting in a moldy, unusable product.

For the actual infusion, my preferred vessel is a canning jar. It is durable, has a lid, and comes in a variety of sizes.

1. Add herbs to the jar and then pour the chosen oil over the herbs until they are completely submerged, leaving a little air space at the top, to help mix things when the jar is inverted.

2. Cover the jar tightly and place in a warm spot. Let the herbs infuse for about two weeks, gently mixing the blend by carefully inverting the jar.

3. For a stronger oil, add a fresh batch of herbs and let infuse for two more weeks. Strain the oil.

4. Add Vitamin E oil at 1 percent of the oil to retard rancidity and prolong shelf life. Keep the strained, infused oil in a cool dark place until needed.

ALL-PURPOSE SALVE

CRACKED SKIN SALVE

This is my favorite salve. I call it my basic first-aid salve. It works really well at soothing and healing minor cuts and scrapes. Simply follow the directions for the Salve Master Recipe on page 56, using infused liquid oil with the following ingredients.

When winter ravages skin, this salve can be a savior. Simply follow the directions for the Salve Master Recipe on page 56, using the following ingredients.

MATERIALS

HERBS	OIL
2 parts calendula blossoms	olive
1 part plantain leaf	
1 part St. John's Wort	ESSENTIAL OILS
	1 part pine needle
	1 part German chamomile
	1 part lavender
	1 part tea tree

MATERIALS

HERBS	OILS
1 part calendula blossoms	equal parts olive and grape seed
1 part lavender blossoms	
1 part comfrey leaf	
1 part plantain leaf	ESSENTIAL OILS
1 part chickweed leaf	lavender or lavandin

BURN SALVE

Minor burns can easily be treated with this lovely salve. Simply follow the directions for the Salve Master Recipe on page 56, using the following ingredients.

MATERIALS

HERBS	OIL
1 part calendula blossoms	olive
1 part plantain leaf	
1 part lavender	ESSENTIAL OIL
	lavender

ITCH SOOTHER

Itches come in many forms. This salve is great for insect bites and other skin conditions that cause itchy skin. Simply follow the directions for the Salve Master Recipe on page 56, using the following ingredients.

MATERIALS

HERBS	OILS
2 parts chickweed	3 parts olive
2 parts chamomile	1 part castor
2 parts calendula blossoms	
1 part thyme	ESSENTIAL OILS
1 part comfrey root	3 parts lavender
1 part marshmallow root	3 parts helichrysum
	1 part peppermint

ROSEBUD SALVE

WOODWORKER'S SALVE

In addition to smelling great, this salve works well for moisturizing elbows and taming fly-away hair. Simply follow the directions for the Salve Master Recipe on page 56, using the following ingredients.

In addition to healing skin, this salve also helps to pull wood or metal splinters to the surface of the skin. Simply follow the directions for the Salve Master Recipe on page 56, using the following ingredients.

MATERIALS

HERBS	OIL
Organic Rose buds	olive
	ESSENTIAL OIL
	rose absolute

MATERIALS

HERBS	OIL
2 parts chickweed	olive oil
2 parts plantain	
2 parts comfrey leaf	**ESSENTIAL OIL**
2 parts rosemary	lavender
pinch bentonite clay	

ICY JOINT & MUSCLE BALM

This soothing balm, made up of natural vegetable oils and beeswax, provides a base for the cooling therapeutic menthol crystals and a blend of therapeutic essential oils.

MATERIALS

yellow beeswax	30 g	25%
cocoa butter	6 g	5%
extra virgin olive oil	26 g	21.5%
almond oil	22 g	18%
castor oil	24 g	20%
menthol crystals	10 g	8%
Vitamin E oil	0.6 g	.5%
eucalyptus dives essential oil	0.7 g	.6%
rosemary essential oil	1 g	.8%
camphor essential oil	0.6 g	.5%
tea tree, Australian essential oil	0.1 g	.1%
double boiler		
digital scale		
wooden stir stick		
1-oz (27 g) tins		
Yield: Four 1 oz (28 g) portions		

1. Warm all the oils, menthol crystals, and beeswax in a double boiler over low-medium heat, stirring until all solids are melted or dissolved.

2. Remove from heat and allow to cool slightly. Add the remaining ingredients.

3. Pour into 1 oz (27 g) tins and allow to cool completely.

I recommend storing it in 1 oz (27 g) containers so they can easily be tossed into a gym bag or carried to the job site where they can provide convenient and instant relief from nagging muscle aches and pain.

VARIATION

If desired, this also works well in large lip balm tubes. Additional beeswax may be needed to achieve the right consistency for stick application.

EMULSIFYING SUGAR SCRUB

I love a good scrub. Although there are oil-based scrubs available, I honestly find that without an emulsifier, they leave a layer of oil on my skin that I feel inclined to scrape off.

As an alternative to oil-based scrubs, try an emulsified scrub, which basically turns into a lotion with the addition of water from the bath or shower, leaving the skin moisturized but not greasy.

NOTE: Although this recipe does not contain water, I have included a preservative to prevent bacteria growth in the scrub. Since the scrub is used around water, the likelihood of water being introduced is very high.

MATERIALS

safflower oil	120 g	12%
sweet almond oil	100 g	10%
cocoa butter	60 g	6%
mango butter	60 g	6%
E-wax NF	40 g	4%
yellow beeswax	7.5 g	.75%
scent	7.5 g	.75%
Optiphen	5 g	.5%
Vitamin E oil	5 g	.5%
sugar, granulated	595 g	59.5%
kitchen scale		
double boiler		
spoon		
thermometer		
small mixing bowl		
hand mixer with whisk attachment or stand mixer		
four 8 oz (227 g) jars with lids		

Yield: Approximatelyt four 8 oz (227 g) jars

1. Weigh safflower oil, sweet almond oil, cocoa butter, mango butter, emulsifying wax, and beeswax and add to a double boiler.

2. Heat until everything is completely melted, stirring occasionally.

3. When the temperature of the melted mixture reaches 120°F (49°C), add Optiphen preservative, Vitamin E oil, and scent blend. Stir.

4. Transfer mixture to a small mixing bowl and place in the refrigerator to cool to around 80°F (27°C).

5. Remove and whip using a hand mixer with whisk attachments or stand mixer. Whisk until it looks like pudding.

6. Add sugar and whisk until well incorporated. Pour into jars and allow to set up.

BEESWAX CREAM

I'll be honest. I was very close to not including any kind of emulsified lotion or cream using beeswax. Why? Because it is a very difficult product to make successfully with the tools that are readily available to the home crafter. Commercial operations use fancy homogenizers to mix the ingredients at very specific temperatures, achieving a product that looks and feels good, is stable, and will not separate over time.

Having said that, there are products that can be made that utilize beeswax as a thickener, instead of relying on the beeswax to help with the emulsification system. This recipe for an all around body cream is a great starting point. It has a nice skin feel, absorbs quickly, and doesn't feel too greasy. To make a more specialized cream, the oils and butters in Phase B can be swapped out to make something that works for the application in mind. If the same consistency is desired, try to keep the percentage of butters and oils the same. Do not change the proportion of emulsifier to wax.

MATERIALS

PHASE A		
distilled water	260 g	56.5%
honey	9 g	2%

Phase B		
E-wax NF	23 g	5%
Stearic Acid	23 g	5%
yellow beeswax	23 g	5%
mango butter	23 g	5%
coconut oil	69 g	15%
lanolin	9 g	2%
Vitamin E oil	14 g	3%

Phase C		
Optiphen	5 g	1%
scent	2 g	.5%

2 heat proof quart size canning jars

double boiler

digital scale

thermometer

immersion blender

four 4 oz (113 g) jars with lids

bowl large enough to use for cold water bath

Yield: Four 4 oz (113 g) jars

BEFORE YOU BEGIN

The heat and hold process referred to in the following instructions simply means that you should heat the mixture to the temperature indicated and hold—or maintain—that temperature for the time provided. This ensures that the water phase remains hot enough for a long enough time to kill off whatever pathogens might be in the water and to completely melt particles in the oil phase.

1. In a pan, combine the ingredients listed in Phase A. In a separate pan, combine the Phase B ingredients.

2. Place the two pans into a double boiler and heat and hold both phase A and Phase B in separate containers at a temperature of 158°F (70°C) for 20 minutes.

3. Pour Phase B into Phase A and combine, using an immersion blender on high. Place the jar in a cold water bath to help cool the mixture down quickly, continuing to blend on high until the mixture reaches a temperature of 104°F (40°C).

4. Add phase C. Continue mixing on low until phase C has been thoroughly incorporated, taking care not to lift the immersion blender up, which will incorporate air. Pour into four 4 oz (113 g) jars.

SOLID NATURAL PERFUMES

I love creating with essential oils and for me, making perfumes is particularly rewarding when I am able to achieve something that is better that the sum of its parts.

My first exposure to solid perfumes was when I toured a perfume house in Grasse, France. I wanted to go home with something fun to remind me of my trip, so I bought a six-pack of solid perfumes. These perfumes ended up being some of my favorite souvenirs from France. I love the tactile and sensuous nature of solid perfumes. When I started creating perfumes myself, solid perfumes were first on my list.

A well-made natural perfume strikes the perfect balance between top notes, heart notes, and base notes. Those who are unfamiliar with 100 percent natural perfumes will be amazed at the complex "story" each perfumes relates. The top notes are the first ones to "speak," and are usually most exuberant. When the top notes have calmed down a bit, the heart notes kick in to continue telling the bulk of the story. The base notes are heard from last, but once they are noticed, the realization hits that they have been there all along. The base notes are the workhorses of the perfume and are responsible for its overall character and staying power. Solid perfumes are to be applied to pulse points and, unlike their spray counterparts, are intended for a discreet audience.

Creating a perfume is not a leisurely afternoon's activity. Perfumes need time to marry and meld and can sometimes take a year or more to finish its magic. The recipes I have included here are best left to marry for a couple months, but they will be ready to use in less than a year.

Although these perfume recipes can be made in an oil or alcohol base, the scent of the natural beeswax in this recipe becomes a unique part of the perfume blend.

I should note that the following recipes will also work with fragrance oils. Fragrance oils can also be blended, but keep in mind that by nature the oils are already a blend of different notes. Before adding fragrance oils to your perfumes, be sure to check the manufacturer's recommended safety levels for skin application.

SOLID PERFUME BASE BALM RECIPE

I like to make up a larger quantity of my Solid Perfume Base Balm Recipe and then just warm the amount I need for the particular container I am filling. There are endless possibilities for containers for these signature perfumes, from small jars to lockets that can be worn around the neck, releasing scent. The recipe for this base is harder than my salve base, since the idea is to pick up just a little bit when the finger is rubbed across the surface.

MATERIALS

liquid carrier oil (jojoba or fractionated coconut oil)	60 g	60.00%
yellow beeswax	40 g	40.00%
heat-proof glass measuring cup		
double boiler		
digital scale		
spoon		
plastic storage container with lid		
Yield: Makes enough to fill approximately 6½ oz (14 g) jars		

1. Add the carrier oil and the beeswax to a small, heat-proof glass measuring cup and warm in a double boiler, taking care not to overheat the mixture.

2. Stir well to combine and pour into a storage container until needed.

SOLID PERFUME MASTER RECIPE

I have written these recipes in parts. When experimenting with the blends, use only drops instead of parts and place into a small bottle or vial for a couple weeks to allow the scents to meld.

MATERIALS

solid perfume base	17 g	85%
essential oil blend	3 g	15%
digital scale		
small deli cup		
wooden stir stick		
microwave		
4–5 lip balm tubes or 2–3 slide tins		

Yield: Makes approximately 4–5 lip balm tubes or 2–3 slide tins

The best way to test the scent of a perfume blend is to use perfume tester strips. Caution, since this is an undiluted essential oil blend, do not try this directly on skin. I use 140-pound acid free watercolor paper and cut it into thin strips that fit into my essential oil blend vial or bottle. Dip the strip into the essential oil blend and sniff it at 5 minutes, 30 minutes, 2 hours, and 5 hours. Write down your impressions of the perfume at the various intervals. Note the changes that the scent undergoes over the course of time. Let the perfume meld for another week and try again.

1. To make a perfume, simply cut a portion of the solid perfume base off the block with a sharp knife.

2. Add the desired amount of solid perfume base to the deli cup and melt in short spurts in the microwave. It should melt easily in 20–30 seconds. Do not overheat.

3. Add the essential oil blend, stir to incorporate, and pour into the perfume container.

4. Let it sit in place until cool. Add the cap and enjoy.

PERFUME ESSENTIAL OIL·BLENDS

Here are a few great blends to try. Feel free to tweak and adjust as desired. Once the perfect blend is achieved on the scent strip, test on skin, since every person's skin chemistry alters the scent slightly. To try the blend on skin, add a drop of the essential oil blend to about 10–15 drops of a carrier oil and apply to the skin. I hope you find one that you like!

BLEND 1

TOP NOTES	Fennel EO*	2 parts
	Sweet Orange EO	5 parts
HEART NOTE	Rose absolute	3 parts
BASE NOTE	Myrrh EO	2 parts

BLEND 2

TOP NOTES	Tangerine EO	6 parts
	Bergamot EO	6 parts
	Lavender EO	9 parts
HEART NOTES	Clary Sage EO	15 parts
	Nutmeg EO	8 parts
	Ylang ylang EO	11 parts
BASE NOTE	Patchouli EO	14 parts

BLEND 3

TOP NOTE	Lavender EO	20 parts
HEART NOTE	Chamomile EO	2 parts
BASE NOTE	Benzoin EO	1 part

BLEND 4

TOP NOTE	Bergamot Mint EO	5 parts
HEART NOTES	Spruce EO	10 parts
	Cypress EO	5 parts
BASE NOTES	Benzoin resin	3 parts
	Cedarwood EO	4 parts
	Vetiver EO	3 parts

BLEND 5

TOP NOTE	Sweet Orange EO	4 parts
HEART NOTES	Coriander EO	3 parts
	Juniper EO	3 parts
BASE NOTE	Frankincense EO	1 part

BLEND 6

TOP NOTE	Bergamot EO	3 parts
HEART NOTES	Palmarosa EO	2 parts
	Rose Absolute	2 parts
BASE NOTE	Sandalwood EO	4 parts

BLEND 7

TOP NOTES	Bergamot EO	2 parts
	Lavender EO	2 parts
HEART NOTE	Clary sage EO	2 parts
BASE NOTES	Helichrysum EO	4 parts
	Vanilla Absolute	1 part

*Essential Oil

CHAPTER 5
SOAPS

Soap, which has been around forever, is one of life's basic necessities. While the actual origin of soap is unclear, we know that for the past 5,000 years, people have been combining alkaline salt with fats to make soap. The Sumerians created a slurry by combining animal and vegetable fats with ashes and boiling the mixture. Ancient Egyptians later wrote about soap recipes that called for mixtures of fats and alkaline salts. The Romans were using bar soap in their baths by 200 CE.

The main difference between the soaps we use today and what people used way back when is that our soaps are reliably gentle on the skin and consistent from one batch to the next. This is because now we know about the chemical reaction that takes place to create soap and we know how to create lye—a main ingredient in soap—that is of a known purity, which ensures results that can be easily predicted and duplicated. We also have modern tools such as digital scales and immersion blenders that help with the task, but the essentials have not changed for the past 5,000 years.

FAT + LYE = SOAP

Without lye, there is no soap. Lye is required to create the chemical reaction that produces soap. When a soap recipe is calculated correctly, no lye remains in the final product. All the lye is converted to a salt.

Why put beeswax in soap? Beeswax helps make the soap emollient and creates a hard bar that lasts longer in the shower. The biggest mistake that most beginners make is adding too much beeswax, which creates a bar that feels rubbery and a bit gummy. But adding just a little bit of beeswax, say 1 to 2 percent of added oils, improves the bar considerably.

Since beeswax is not an oil, not all of the wax will react with the lye. Beeswax contains about 50 percent unsaponifiables. Unsaponifiables are the substances in beeswax that do not react with the lye. The unsaponifiables may be substances that help decrease the trace time in recipes using beeswax, but are also probably the substances that make a soap nice for the skin.

I consider using beeswax in soap to be an advanced soap-making technique. The main reason is that beeswax has a melting temperature of at least 145°F (63°C), while most soaps are made at a temperature that is quite a bit cooler than that, at say 100°F (38°C). The oils and lye need to be combined at a higher temperature to ensure that the beeswax remains melted. The higher temperatures will make the

soap saponify more quickly, which could result in a failed batch if you're not careful.

For beginning soap makers, I have included a beeswax-free starter recipe that can be made easily with oils found at most grocery stores. I strongly urge people who have not made soap before to make the starter recipe first—to become familiar with the process and gain an understanding of the different stages the soap goes through. Once the basic soap-making techniques are mastered, the cold process or hot process soap recipes that follow will be less daunting and you're more likely to make a successful batch of soap.

Each oil used in soap making has a Saponification Ratio value, also known as the SAP value, which is the amount of lye needed to convert that oil into soap. Since each oil has a different value, do not make any oil substitutions without recalculating the amount of lye needed for the oil you are considering. The amount of lye needed can be calculated by hand or using an online calculator. I recommend running every recipe through a lye calculator, regardless of the source. The recipe could have inadvertent errors like transposed numbers or incorrect units. It is also nice to have

a paper copy to check off ingredients as they are added and add notes on scent or color. Using the lye calculator also allows for resizing a recipe to fit a particular mold. There are free web-based lye calculators that will calculate the amount of lye needed and allow the user to print out the recipe. Some calculators have predictive values for hardness and lather that are useful when creating recipes from scratch or tweaking recipes to fit the ingredients on hand.

Although it is possible to make a soap with just one oil, there are reasons that most soap makers use more than one. Each oil has a different breakdown of essential fatty acids that bring different properties to the soap. Coconut oil is really high in lauric acid, yielding a soap that makes loads of bubbles, will dissolve really quickly, and is very cleansing (too harsh for most people if used in high proportion). Olive oil, on the other hand, which is loaded with oleic acid, makes a soap that has more of a creamy lather (that almost feels slimy if not properly aged), is super gentle on the skin and once fully cured, is hard as a brick. Most of the other oils will fall somewhere between coconut and olive oil. In the ingredient glossary, I have included profiles for all the oils I use in this book plus a couple more that might make nice substitutions.

SOAP-MAKING EQUIPMENT

Soap-making equipment only has a couple requirements. First, if using metal, ONLY use stainless steel. The lye will react with other metals. Second, never use glass to mix the lye. Over time, the lye will create minute cracks and scrapes in the glass, eventually causing it to shatter, potentially spilling hot, caustic lye water all over the place.

Use a disposable paper plate or bowl to measure the dry lye. This makes clean-up easy. Lye crystals can have quite a bit of static electricity and the individual beads can escape. Take care when measuring and give all surfaces a quick wipe down once the lye is in water.

I recommend using an immersion blender (A) to mix the lye with the oils. Immersion blenders are available at big box stores at reasonable prices and they help to ensure that the soap batter is mixed properly. The immersion blender has a high shear blade that mixes the soap easily and reliably. A hand whisk (B) or spoon can also be used, but it will take quite a bit longer.

For mixing the soap, I like to use stainless steel stock pots, bowls (C) or plastic buckets. A small, flat-bottomed mop bucket works really well, since it is durable enough to withstand some warmer temperatures and is not too tall, allowing the soap maker to use an immersion blender or whisk with ease. A mop bucket also has a handle, which is useful for transferring the soap batter to the mold.

Soap molds can be as simple as a lined cardboard box (D) or as complicated as an ornate silicone mold, with everything in between. The recipes in this book are

sized for a regular bread loaf pan (I like silicone, which doesn't need to be lined), which yields approximately eight bars of soap. For lining soap molds, I prefer freezer paper. Line the molds with the shiny, coated side face up, resting against the soap. To line a mold or box, cut two pieces of freezer paper and lay them in the box in opposite directions, making a nice crisp corner where the bottom meets the sides and leaving the extra folded over the sides of the box. There will be a slight gap at the corners, where the two pieces of freezer paper meet, but I have not found that to be a problem as long as the soap is poured into the mold at the proper time.

A digital scale (E) is imperative for these recipes, as all ingredients are measured in grams. Reasonably priced scales can be found at any big-box store. Measure solid ingredients on a piece of freezer paper and liquid ingredients in plastic deli containers (always remember to zero out the weight of the container before adding ingredients).

PROTECTIVE GEAR

Lye requires several items be used for personal and environmental safety. In crystal form, lye can be annoying, but it is relatively harmless. It can develop a static charge that causes the lye to end up far from its original container. Once it has been dissolved in water, it is much more caustic, but less likely to create problems since the liquid is easier to manipulate safely.

Clothing should be the first level of protection. Wear a long-sleeve shirt, long pants, and closed-toe shoes. On top of that, gloves and goggles help keep your hands and eyes protected from errant lye crystals and from raw soap mix. Next, when the lye is being added to water, the reaction gives off some fumes that can be irritating and are better not breathed in. I like to mix the lye into water outside where the fumes aren't a problem, but for those who prefer to do this in their soaping space, I recommend the use of a respirator. It is not needed for the rest of the soap-making process.

STARTER RECIPE (NO BEESWAX OR HONEY)

MATERIALS

olive oil (I prefer pomace grade)	312 g	36.7%
coconut oil (76°F [24°C])	227 g	26.7%
Crisco ("palm" recipe, see note)	283 g	33.3%
castor oil	28 g	3.3%
distilled water	323 g	
lye (NaOH— sodium hydroxide)	117 g	
fragrance (if desired)	.43 g	

disposable paper bowl

container for lye (see Soap-Making Equipment, page 73)

liquid measuring cup

stock pot or microwave-safe container

stainless steel stock pot or flat-bottomed mop bucket

stainless steel whisk or immersion blender

digital scale

large plastic cooking spoon

chef's knife

mold (silicone bread loaf pan preferred, but any mold will do)

freezer paper (if loaf pan is not silicone)

piece of cardboard, cut to the size of the loaf pan

Yield: Approximately eight 4 oz (113 g) bars

BEFORE YOU BEGIN

The first thing I like to do when I make soap is to gather all my ingredients together and put them on my table in the order in which they appear in the recipe. I do this for two reasons. First, I can check to make sure that I have enough of all my ingredients to actually make the soap. It's better to realize that there is no castor oil before making soap, rather than right in the middle of soap making. Secondly, I can grab the oils in order and check them off on my printed recipe. It helps to ensure that I measure out the correct amounts.

NOTE: I specifically call for Crisco instead of generic vegetable shortening, since Crisco's Saponification Ratio (SAP) is known by the lye calculators and the proportions of oils in the vegetable shortening blend may be different with other brands. Also, Crisco changed their formula several years back and version commonly found on store shelves is made with palm oil. Check the ingredients listing on the Crisco package to make sure it includes palm oil.

1. Always make the lye solution first, as it requires cooling time. Measure the lye into a disposable paper bowl and set aside.

2. Measure out the correct amount of distilled water and pour into the lye-safe container. Place the container on a high heat-resistant surface and add the dry lye crystals to the water (never the other way around). Stir with a plastic cooking spoon until all the lye is completely dissolved. Set the lye mixture aside to cool.

3. Prepare the mold, lining with freezer paper with the shiny side face up.

4. Heat all the solid oils in a stock pot on the stove top or in the microwave in a heat-safe container. Once melted, pour the oils into the mixing container, either a stock pot or a flat-bottomed mop bucket. Add the liquid oils and stir to mix.

5. Check the temperatures of the lye and the oils. Ideally, both should be around 100°F (38°C). If the temperatures are higher, let it cool down a bit more.

NOTE: When using honey, beeswax, or any kind of milk, those ingredients can cause the soap to get hotter than normal and it may not be necessary to cover the mold.

9. Allow the soap to sit for several hours. Since soap making is an exothermic chemical reaction, the soap will heat up in the next couple hours and go from what looks like a nice creamy soap into something resembling petroleum jelly and it will be HOT. That's what is called the "gel" phase. While it's not completely necessary for a batch of soap to go through gel, I think it makes for a better, more consistent end product.

10. After about twenty-four hours, the soap should be cool, relatively hard, and ready to unmold and slice. If it still seems a bit soft, leave it in the mold a bit longer and check it again after another day or so. Once it seems hard enough, use the chef's knife to cut the soap into individual bars, keeping in mind that there is still quite a bit of water in the soap that will evaporate over time, causing the bar to shrink a bit. Set the cut bars on some freezer paper with a bit of air space between them and set them somewhere out of the way to dry and cure. Cure time is at least one month.

6. To make the soap, pour the lye water into the oils and mix with the immersion blender (A). Notice how the mixture changes from transparent to milky. Continue mixing until the soap reaches a stage called "trace." You will know that you are at the trace stage when you pull the immersion blender out of the soap mixture and it leaves a visible trail in the top of the soap (B). I like to take my soap to a medium to heavy trace, which is almost the consistency of a soft pudding. Depending on the temperature of the oil and lye and the speed of the immersion blender, the actual mixing portion should not take that long, maybe five to ten minutes.

7. If this is your first time making soap, I recommend not adding any color or scent. But once the recipe is more familiar, this would be the time to add those ingredients.

8. Pour the soap batter into the prepared mold. Scrape out the soap pot and tap the soap mold on the counter a couple times to make sure there are no air pockets. Smooth out the top and cover the mold with a piece of cardboard to hold in some of the heat.

HONEY AND BEESWAX SOAP (COLD PROCESS RECIPE)

Once the basic recipe has been mastered, it is time to add some goodies such as beeswax and honey. The beeswax will make a harder bar and the honey will boost lather and provide some moisturizing benefits.

MATERIALS

olive oil	358 g	44.8%
coconut oil (76°F [24°C])	225 g	28.1%
palm oil	177 g	22.2%
castor oil	32 g	4.0%
beeswax	7.2 g	0.9%
distilled water	276 g	9 oz
lye (NaOH— sodium hydroxide)	111 g	
honey	1 tbsp	
fragrance (if desired)	2 tbsp	
disposable paper bowl		
container for lye (see Soap-Making Equipment p73)		
large plastic spoon or high heat spatula		
stainless steel pot or microwave-safe container for oils		
stainless steel whisk or immersion blender		
digital scale		
mold (silicone bread loaf pan preferred, but any mold will do)		
freezer paper to line mold		

Yield: Approximately eight 4 oz (113 g) bars

1. Gather all the ingredients together and arrange them on the table in the order in which they appear in the recipe.

2. Always make the lye solution first, as it requires cooling time. Measure the lye into a disposable paper bowl and set aside.

3. Measure out 5 oz (140 g) of distilled water and pour into the lye-safe container (see Soap-Making Equipment, page 73). Place the container on heat resistant surface and add the dry lye crystals to the water (never the other way around). Stir until all the lye is completely dissolved. Set the lye mixture aside to cool.

4. In a microwave-safe container, add the honey to the remaining 4 oz (115 g) of water and stir to incorporate. Microwave for a couple of seconds at a time until the honey is completely dissolved. Set aside.

5. Prepare the mold, lining with freezer paper with the shiny side face up.

6. Heat all of the solid oils and beeswax in a stainless steel pot on the stove top or in a microwave-safe container in the microwave. Once melted, pour the melted solid oils into the mixing container and add the liquid oils, stirring well to ensure it is all mixed.

7. Check the temperature of both the lye and the oils. To keep the beeswax from hardening, the ideal temperature of the oils will need to be around 120°F (49°C). The goal is to have the lye right around that temperature as well.

8. Add the reserved honey water to the lye water. It will probably turn colors; mine usually turns some sort of pinkish hue. That's normal.

9. Pour the lye water into the oils and mix with the immersion blender. Once it is emulsified, but not yet at trace, add the fragrance if desired. Keep mixing until it gets to "trace." For this soap, I recommend mixing until it is a light to medium trace. Once trace is achieved, work quickly to get the soap into the mold, as it may solidify quickly.

10. Pour the soap batter into the prepared mold, taking care to scrape all of the soap residue out of the pot. Tap the soap mold on the counter a couple of times to remove any air pockets. Smooth out the top and cover the mold with a piece of cardboard to hold in some of the heat.

NOTE: When using honey, beeswax, or any kind of milk, these ingredients can cause the soap to get hotter than normal and it may not be necessary to cover the mold.

11. After about twenty-four hours, the soap should be cool, relatively hard, and ready to unmold and slice. If it still seems a bit soft, leave it in the mold and check it again after another day or so. Once the soap seems hard enough, cut it into individual bars. Place the cut bars on freezer paper with space between them and set aside for about a month to dry and cure, rotating them occasionally so that all sides dry evenly.

12. I like to store my soaps in a cool dry place until needed. As they age, they will continue to lose water, making them longer lasting in the shower and more mild, so older soaps are a good thing to have around. If they are to be given as gifts, a paper cigar band, paper box, or muslin bag are best for packaging, since they allow the soap to breathe.

HONEY, OATS, AND BEESWAX SOAP (HOT PROCESS RECIPE)

Hot process soap is soap that is cooked through the saponification stage. I like to use a slow cooker with a "warm" setting for this recipe. Make sure that there is plenty of room, approximately double the size of the soap volume, since it may "fluff up" a bit during cooking. A 5–6 quart (5–6 L) slow cooker should be a good size. This recipe also has some oats for soothing and scrubbiness. I like to run mine through a coffee grinder or food processor first to chop the oats into smaller pieces.

MATERIALS

olive oil	370 g	44.8%
coconut oil (76°F [24°C])	232 g	28.1%
palm oil	183 g	22.2%
castor oil	33 g	4.0%
beeswax	7.5 g	0.9%
distilled water	285 g	10 oz
lye (NaOH— sodium hydroxide)	115 g	
honey	1 tbsp.	
oats	4 tbsp	
fragrance (if desired)	2 tbsp	
disposable paper bowl		
container for lye (see Soap-Making Equipment, page 73)		
large plastic spoon or high heat spatula		
slow cooker		
stainless steel whisk or immersion blender		
digital scale		
mold (silicone bread loaf pan preferred, but any mold will do)		
freezer paper to line mold (if not a silicone mold)		

Yield: Approximately eight 4 oz (113 g) bars

1. Gather all the ingredients together and arrange them on the table in the order in which they appear in the recipe.

2. Prepare the mold, lining with freezer paper with the shiny side face up.

3. Measure the lye into a disposable paper bowl and set aside.

4. Measure out the distilled water and pour into the lye-safe container (see Soap-Making Equipment, page 73). Place the container on heat-resistant surface and add the dry lye crystals to the water (never the other way around). Stir until all the lye is completely dissolved. With hot process, there is no need to wait until the lye and the oils are the same temperature.

5. Melt the oils in the slow cooker on low. Once they are completely melted, add the lye mixture and blend with the immersion blender. Keep blending until it gets to "trace."

6. With the heat on low, continue to stir occasionally using the whisk as the soap goes through various stages. Early on, it may separate. Later on it may bubble up and threaten to boil over. If that happens, turn off the heat temporarily and whisk a little harder to deflate the bubbles. Once it has settled down again, put it back on the heat and continue the cooking process. Once it is at a stage where it resembles petroleum jelly, which usually takes about 30–40 minutes of cooking time, it is almost done. Be careful not to overcook the soap. Take it to just the petroleum jelly stage and then remove from heat.

7. Add the honey, oats, and fragrance, if desired, and stir well before transferring the soap batter into the prepared mold. Tap the soap mold on the counter a couple times to make sure there are no air pockets. Smooth out the top and set aside to cool.

8. When completely cool, cut the soap into individual bars. Place the bars on freezer paper with space between them and set them aside to dry for at least two weeks. Although with the hot process soaps they are fully saponified and useable, there is still extra water in the bars that should be allowed to evaporate.

CHAPTER 6
HOME PRODUCTS

Beeswax products are a great addition to the home as well, ranging from decorative to completely useful. Ever since Man discovered beeswax, he has been using it to make his life better. He discovered that wax made surfaces more water resistant and more durable and it could be molded into a variety of shapes. It made wood nice and smooth and sealed it from the elements. It conditioned leather, making it more water resistant and supple. It strengthened thread and string, so items made with it lasted longer.

BEESWAX ORNAMENTS

Beeswax ornaments are a great way to scent and beautify any room, especially during the holidays. They are easy to make and last a long time.

MATERIALS

1 lb (425 g) beeswax

fragrance oil (if desired)

string or ribbon for hanging (6' [1.8 m] cut into 8–10" [20.5–25.5 cm] lengths)

pouring pitcher

double boiler

thermometer

candy molds

silicone mold release (if using metal molds)

toothpick

Yield: 10–12 ornaments

1. If you are using metal candy molds, spray them with silicone mold release.

2. Melt the beeswax in the pouring pitcher using the double boiler. Once the wax is melted, check the temperature. If it's hotter than 170°F (77°C), let it cool to 170°F (77°C). Add the fragrance in the amount suggested by the manufacturer, stir to incorporate, and pour into the candy molds.

NOTE: Other molds can be used, such as silicone muffin molds, which now come in a variety of seasonal shapes. Those work well, but I would suggest not filling them all the way. Half way is more than enough.

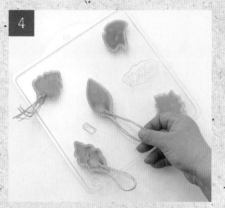

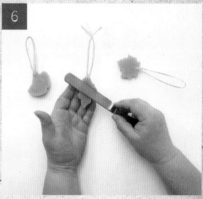

3. Take each 8–10 inch (20.5–25.5 cm) length of string and fold it in half. One end will be embedded into the wax. I like putting the looped end into the wax so I can tie the ornament onto tree branches or other locations, but either end will work.

4. Once the wax begins to skin over on the mold side (but not on top), add the string, pushing it into position with a toothpick to embed it in the wax. Make sure the string is centered and completely surrounded by wax, not just sitting on top of it.

5. Allow to cool in the mold. Once it is completely solid, invert the mold on the tabletop and tap gently to release. If any ornaments stick to the molds, place the mold in the freezer for 5–10 minutes and try again.

6. Clean up the ornaments on the backside by warming a butter knife over a flame and using it to melt away the rough edges.

7. Hang the ornaments and enjoy. If the ornaments get dusty, wipe them down with rubbing alcohol and a soft toothbrush. Allow to dry. To revive the scent, warm the ornament slightly with a blow dryer.

BEESWAX DIPPED LEAVES

I love fall. Here in Wisconsin we usually have a very colorful fall that is always more fleeting than I would like. This project is a perfect way to preserve some of that fall color to enjoy year round. For this small investment in beeswax, you can make a tree's worth of leaves, if you desire.

MATERIALS

1 lb (425 g) beeswax

dried leaves (colorful, pliable leaves work best)

fragrance (if desired)

small electric skillet

waxed paper or freezer paper

1. First prepare the leaves. Sort through all the leaves and select the best ones for this project. They need to be dry, so put them in a warm oven (about 200°F [93°C] for about a half hour) to dry if necessary.

2. Melt the beeswax in the electric skillet, taking care not to overheat the wax. Once the wax is melted, add the fragrance in the amount suggested by the manufacturer and stir to incorporate. The wax level should be high enough to easily immerse the leaves into the hot wax.

3. Working with one leaf at a time, hold onto the stem and dip the leaf into the hot wax. Use a wooden chopstick or toothpick to push the entire leaf under the wax. Remove and set aside on the waxed paper to cool.

4. Place the finished leaves in a decorative bowl or string them on a fishing line to create a garland.

BEESWAX PINECONE FIRESTARTERS

Start a fire in style with these pinecone firestarters. They are stylish enough to beautify a space and fill the room with light, pleasant scent. They require quite a bit of wax but it doesn't have to be the super high-quality wax used in candle making.

MATERIALS

2 lbs (850 g) beeswax for pinecones (lower grade wax is fine here)

5 lbs (2.3 kg) beeswax for dipping container

12 pinecones (see note)

candle wicking (I prefer square braid 2/0 cotton wicking)

fragrance or essential oil, optional

double boiler

thermometer

wooden stir stick

freezer paper

Yield: Approximately 12 medium-size pinecones

NOTE: Pinecones that aren't completely dry may explode once they are in the fire, so to be safe, make sure that the pinecones are completely dry. I like to place them on a baking sheet and "bake" them for an hour or so at the lowest oven setting. If the oven doesn't go real low (mine goes down to 150°F [65°C] which is perfect), heat the oven on the lowest setting and then turn it off and add the pine cones. Leave them in the oven until cool. Now they are ready to be made into firestarters.

1. In the double boiler, heat the wax to approximately 190–200°F (88–93°C).

2. While the wax is heating, tie an 8–10" (20.5–25.5 cm) piece of wicking around the pinecone close to the top, leaving a long tail and cutting the other end close to the knot. Thread the long end up, close to the middle of the pinecone at the top. This will ultimately be the wick for lighting the pinecone, but it's also useful for dipping the cone into the wax. Prepare all the pinecones the same way and set aside.

3. Once the wax is hot enough, add the scent, if desired, and give it a quick stir to mix. Holding the pinecone by the wick, dip it into the wax and leave it there until all the bubbles stop coming up—probably about 10 seconds or so. The idea is to get a thin coating on the pinecone. When it first goes into the wax, it will be cold and more wax will adhere, but as it warms up some of the wax will melt off again. At this point, pull the cone out of the wax and set it on the freezer paper to cool. Continue with the remaining pinecones.

4. To use, simply light the wick and place in the fire. Enjoy!

BEESWAX LUMINARIES

These little gems are the perfect way to create mood lighting inside or in outdoor entertaining spaces—and they make use of handmade beeswax tealights or votive candles. This is a fun project to do with older children.

MATERIALS

1 lb (425 g) or more beeswax for
luminaries

5 lbs (23 kg) for dipping container

double boiler for wax

balloons (high quality round ones
work best)

cold water

Yield: 3–6 luminaries

1. Melt the beeswax in the double boiler.
Use a container that is large enough to be
able to fit the balloons in with at least a
little wiggle room.

2. Fill a balloon with cold water, paying
attention to the shape that the water
creates as it fills the balloon. Once the
desired shape is attained, tie off the
balloon. Make sure that the balloon is not
stretched too much or it may burst when it
comes in contact with the hot wax. When
determining the size and shape of the
votive, also consider the heat that will be
given off by the candle. It is better to make
the luminary slightly larger than needed
rather than risking having it melt because it
was made too small.

3. Holding the balloon by the tie-off knot,
dip it part of the way into the melted
beeswax and pull it back out. Set the
balloon lightly on a table to create a flat
spot on the bottom that will help to keep
the luminary stable. The cold water in the
balloon will harden the wax quickly but the
shape established early on will determine
the final shape, so work fast.

4. Dip the balloon and set it on the table
to harden several more times until the
luminary has the desired thickness. I like
to do mine five to six times. Experiment
with dipping the balloon at a different angle

each time, keeping in mind that those
variations will show once there is a candle
inside.

5. Drain the water out of the balloon by
either undoing the knot or by popping the
balloon. The balloon will peel right off
the wax. If desired, the luminary can be
left like this. However, I like to clean up
the luminary a bit more by using my hot
plate to melt a flat area on the bottom
and smoothing out the edges. Simply turn
the luminary upside down and lightly run
its top edge along the surface of the hot
plate. Add a beeswax tealight and enjoy the
ambience.

CUTTING BOARD CONDITIONER

Wooden cooking tools such as cutting boards, wooden spoons, and rolling pins need occasional care, and since they will be in contact with food, it's necessary to use a food grade product. I like to use a food-grade mineral oil, since I know that it will last indefinitely and does not get gummy over time. For cutting boards that see a lot of use, walnut oil is an option as well.

MATERIALS

yellow beeswax	70 g	20.6%
food-grade mineral oil	270 g	79.4%
double boiler		
digital scale		
spoon or stir stick		
wide-mouth jar or tin with lid		
Yield: 5 oz		

1. Heat the beeswax and mineral oil together in a double boiler until melted. Mix well and pour into wide-mouth jars.

2. To use, simply scoop out some of the conditioner with a rag or your fingers and rub it all over the wood. Let it sit overnight and buff it with a soft cloth the next morning.

SOLID WOOD POLISH

Keep wood looking beautiful with this age-old recipe.

MATERIALS

yellow beeswax	75 g	33.33%
boiled linseed oil	75 g	33.33%
turpentine	75 g	33.33%
double boiler		
digital scale		
spoon or stir stick		
wide-mouth jar or tin with lid		
Yield: Approximately 8 fluid ounces (235 ml)		

Combine all three ingredients in a double boiler, stirring occasionally. When it is all melted, transfer to a wide-mouth jar or tin and allow to cool.

WOOD CONDITIONER CREAM

This recipe is perfect for nourishing the wood. It cleans and protects in one step.

MATERIALS

yellow beeswax	150 g	71.5%
turpentine	60 g	28.5%
double boiler		
spoon or stir stick		
wide-mouth jar with lid		

Yield: Approximately 8 fluid ounces (235 ml)

1. Heat the wax in a double boiler until melted and remove from heat.

2. Stir in the turpentine. Mix well and pour into a wide-mouth jar.

BEESWAX FURNITURE POLISH

Beeswax furniture polish is considered the ultimate in wood care due in large part to the soft, satin shine it gives to the wood. The touch of carnauba wax in this polish results in a harder, shinier finish.

MATERIALS

beeswax, yellow	43 g	10%
carnauba wax	10 g	2.4
turpentine	372 g	87.6%
double boiler		
digital scale		
spoon or stir stick		
wide-mouth jars or bottles with lids		
Yield: Makes approximately 16 fluid ounces (475 ml)		

1. Melt the waxes in a double boiler.

2. Remove from heat and stir in the turpentine.

3. Once it is completely mixed, pour into wide-mouth jars.

4. Apply the polish with a clean cloth and rub in small circles. Turn the cloth as it becomes dirty. Allow the polish to dry, then buff with a clean cloth. If more than one coat is desired, wait two days between applications.

WAXED FABRIC

Applying beeswax to fabric has long been a method of water-proofing something that normally is anything but waterproof.

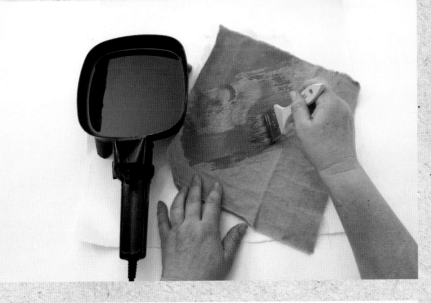

Almost any smooth, natural-fiber fabric can be made more water resistant by applying a layer of beeswax, but some fabrics, such as corduroy, just don't lend themselves as readily to this process. Canvas, twill, muslin, and numerous other fabrics work really well and can be turned into a multitude of useful, environmentally friendly tools.

.How best to apply the wax is determined by the object you'd like to waterproof. If you are starting with a flat fabric, the best way is to soak the fabric in beeswax and then removing the excess. This process ensures that the wax penetrates the fibers and creates a film keeping water out. When working with objects that are already made, such as canvas sneakers, the wax will need to be applied to the finished item.

WATERPROOF SNEAKERS

Imagine sloshing through water puddles in sneakers and not getting wet feet. Sounds great, right? Well, I can't promise that the end result will yield completely dry feet, but they will be significantly drier. Unlike plastic rain shoes, the beeswax still allows feet to breathe, and you'll look a lot more stylish!

This technique can be employed for anything already made, not just sneakers.

MATERIALS

3 oz (212.5 g) beeswax

pair of canvas sneakers

2 tall kitchen garbage bags

newspaper

paper towels

small electric skillet

cheap 2" (5 cm) natural bristle brush

heat gun

1. Stuff each of the two kitchen garbage bags with a few sheets of newspaper then stuff the garbage bags inside the shoes. The plastic will keep whatever wax seeps through from sticking to the paper inside the shoe. The paper is there to help the shoe keep the proper shape.

2. Heat the electric skillet to around 200°F (93°C) and add the beeswax. When completely melted, dip the paint brush in the wax and let it warm up for about 30 seconds.

3. Paint the hot wax onto all canvas parts of the shoe, making sure that the critical areas, such as the interface of the sole to the top, is thoroughly soaked. Set the shoe aside and repeat the process with the other shoe.

4. Pull the bags out of the shoes and allow the shoes to cool a bit.

5. To clean off some of the excess wax, stuff a big wad of paper toweling in the shoe and use a heat gun to warm the sloppy section from the outside. The wax will quickly melt and appear to wick away. Move the paper toweling to another area of the shoe that needs work and repeat the same process, replacing the paper towels as needed.

6. Let the shoe cool and check for any missed spots. Touch up the missed sections and try on the shoes. They will feel a bit stiffer until the wax softens up a bit. To rejuvenate the wax coating, hit it quickly with a heat gun. Over time the waterproofing may wear off and require another application. This time only a little wax will be needed, but follow the same process as before.

WAXED COTTON SANDWICH WRAPS

These wraps are perfect in this eco-conscious world we live in. They are simplicity at its finest. They keep a sandwich fresh, keep all the ingredients in place and can serve as a placemat/plate when eating the sandwich. How great is that! In a smaller size they can also be used as a replacement for plastic wrap. Just warm the wrap and mold around the top of the jar or bowl.

MATERIALS

¾ yard (68.6 cm) of medium-weight cotton fabric (I like to use twill fabric for this)

1 lb (425 g) beeswax

electric skillet

tongs

paper towels

heat gun

Yield: Approximately 8 wraps

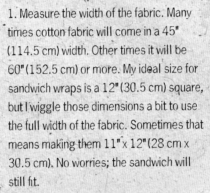

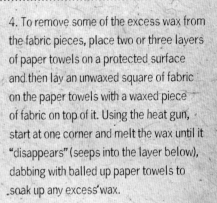

1. Measure the width of the fabric. Many times cotton fabric will come in a 45" (114.5 cm) width. Other times it will be 60" (152.5 cm) or more. My ideal size for sandwich wraps is a 12" (30.5 cm) square, but I wiggle those dimensions a bit to use the full width of the fabric. Sometimes that means making them 11" x 12" (28 cm x 30.5 cm). No worries; the sandwich will still fit.

2. Melt the beeswax in the electric skillet. Once it is melted, lay the wrap in the hot wax and wait for the bubbles to disappear, about ten seconds.

3. Using the tongs, pull the fabric out of the wax and let as much of the wax drip off as possible. As the wax cools, try to straighten out the fabric so that its not stuck together weirdly. There will still be quite a bit of wax on the fabric. Set it aside for now and repeat with another two or three pieces of fabric.

4. To remove some of the excess wax from the fabric pieces, place two or three layers of paper towels on a protected surface and then lay an unwaxed square of fabric on the paper towels with a waxed piece of fabric on top of it. Using the heat gun, start at one corner and melt the wax until it "disappears" (seeps into the layer below), dabbing with balled up paper towels to soak up any excess wax.

5. Continue running the heat gun over the rest of the top fabric piece. Lift off the top piece and check to see how thoroughly the lower piece is now waxed. Most times it just needs a little touch up to make sure it is all evenly waxed. Set these two aside and repeat with the remaining wraps.

6. While the wraps are still warm, make sure that they are completely square. If not, gently tug at the corners to pull them back into shape. Once the wraps are cool they are ready to use, but will be quite stiff. As they are used they will soften up and become more malleable.

7. To clean, wipe down with a damp cloth. If needed, wash wraps in cool water by hand.

BEADING AND SEWING

The thread or string used in jewelry or other beading projects needs to be thin enough to go through the hole of tiny beads and strong enough to withstand all the abuse the wearer will put it through. The same goes for sewing. Beeswax helps strengthen and condition the thread. It prevents fraying and makes beading and sewing easier. It also reduces tangling and knotting, resulting in a nicer, more professional looking product.

THREAD CONDITIONER

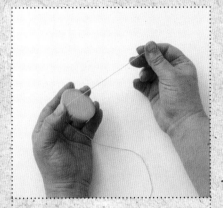

MATERIALS

4 oz (106 g) beeswax

double boiler or small electric skillet

Nonstick muffin tin or beeswax mold

1. Melt the wax in the double boiler, taking care not to overheat the wax.

2. Pour the wax into the mold. I usually fill them about ½–1" (1.3–2.5 cm) deep. The block can be a bit clunky if the wax is poured too deep.

3. Allow the wax to cool completely then pop the wax out of the mold.

4. One last thing I like to do before using the block of beeswax is cut a small groove into the edge of the block with a warm butter knife. Simply heat the edge of the knife briefly over a flame and run the knife along the edge of the wax block. The groove doesn't need to be very deep, but it helps keep the thread in place the first couple times it is used.

5. To use the thread conditioner for beading, take the length of thread or string and lay one end over the notch. Place your thumb over the thread and gently pull the thread with the other hand. Run it through a couple more times, making sure that the whole thread is waxed.

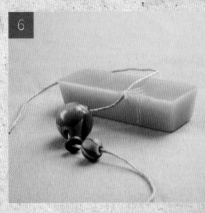

6. Next, pinch one end of the thread between your thumb and forefinger and pull the thread, gently smoothing it. Do this a couple more times and then it is ready to use.

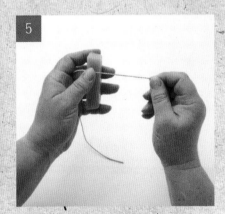

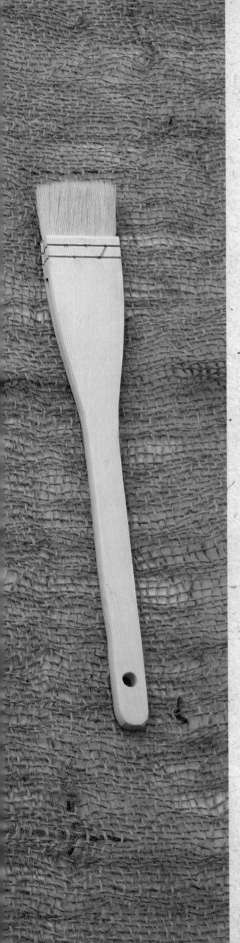

CHAPTER 7
BEESWAX ART

Beeswax is a great medium for creating incredibly diverse art pieces. There are two distinct techniques that I will touch on in this section: batik and encaustic. These two techniques provide fertile ground for bringing out the creative spirit in everyone, and with a few safety precautions, are a lot of fun to do. There are some specialized tools that make it easier to achieve, but none of them are too exotic.

BATIK

Batik is an ancient art form that dates back at least 2,000 years. It is essentially a resist technique, which means that color is being protected from additional dye by being covered with wax. Each color builds on the sum of the previous colors. It is an art form that can takes years to perfect, a vision for the end product, and patience to see the piece through all the steps needed.

Much as I would like to call myself an artist, I am not. But I do like to do artful things and this is right up my alley. I like that batik can be as intricate or simplistic as time and skill will allow. The two projects included in this section are perfect for beginners and they create items that can be used once completed.

Basic hardware and craft-store tools are all that is required, but there are some specialized tools that can be a lot of fun to play with. One that I like to use is a tjanting, which is a stylus that holds molten wax and dispenses it through a narrow tip. It is held like a pen or butter knife. The art-supply store in my area had a set of tjanting tools in small, medium, and large, which was perfect since I had no idea which size I needed to make my vision a reality.

BEESWAX/PARAFFIN RATIO

The wax is generally a blend of beeswax and paraffin. Depending on the ratios of the two, different effects can be achieved. The more paraffin the more the wax will crackle and let some dye through. Personally, I love this effect, so my blend will do a bit of that. Adjust the proportion of the wax blend as desired—more paraffin, more crackle; more beeswax, less crackle.

Also, I encourage you to learn from my mistake and purchase new paraffin wax for batik, rather than recycling used candles. Although candles may seem like they are all paraffin, they may contain oils and other fillers. The oils will not wash out easily and may leave a permanent grease stain, in addition to making the wax more pliable, which is counter to why the paraffin is added in the first place.

BEST DYE TECHNIQUES

For batik, I like to use specially designed cold-water dyes that do not require heat. The grocery store dyes should work as long as they are prepared according to the directions and then allowed to cool. If the dye bath is too warm, it will melt the wax. Wetting the fabric with cold water first will allow the fabric to absorb the dye more evenly. Leave it in the dye bath until the desired tint has been achieved. I recommend using fabric scraps to test the dyeing time. Have several that can be pulled out at regular intervals, given a quick rinse and tossed in the dryer to see how that color looks dry. Once the dry sample is the desired color, remove the batik from the dye bath and rinse with cold water until the water is relatively clear. This is especially important with batik, since any "free" dye remaining in the fabric may tint the areas where there is wax.

REMOVING THE WAX

There are two ways the wax can be removed: boiling it in hot water or ironing it with paper towels. When deciding which route to go, I would suggest using the dyed sample fabric to check the dye's colorfastness. The easier, cleaner way to remove the wax is by putting the whole piece in a stockpot and boiling it. The wax floats to the top and can be skimmed off when the pot cools. The other method is to use a hot iron and melt all the wax into some paper towels. If going with the ironing method, I would suggest multiple layers of newspaper to protect the table surface, then several layers of paper toweling then the batik piece and then another couple layers of paper towels. Keep replacing the paper towels until all the wax is removed.

BATIK NAPKINS

For this project, I decided to use a stamp to apply the wax. I found some old rosette molds tucked away and thought they might work really well, and they do! But it doesn't have to be a rosette mold. Metal cookie cutters work well as do some kitchen implements such as the old-fashioned potato masher. The important points to keep in mind, is that they need to be able to take the heat and they need to be able to be picked up and positioned on the napkin. With cookie cutters, try clipping a clothespin to the edge. For this project, we will only use one dye bath.

MATERIALS

cotton napkins

beeswax

paraffin wax

small electric skillet

metal stamp of choice (cookie cutter, rosette mold)

cold water dye

dye vat (I like to use a stainless steel stock pot, but a plastic bucket should work as well)

clothesline and clothespins

embroidery hoop large enough to hold most, if not all, of the napkin

iron and ironing board, optional

stainless steel stock pot, optional

I will be using 60 percent beeswax, 40 percent paraffin wax blend for this project. How much is needed will depend on the size of the electric skillet and the tool being used. With my rosette molds, I found that the wax only needed to be ½" (1.3 cm) deep.

1. To prepare the napkins, soak the napkins for an hour or so in a solution of water with some detergent and washing soda added. Rinse well and dry.

2. To prepare the wax, add the beeswax and paraffin to the electric skillet and slowly warm until it is melted. Set the skillet temperature to between 180°F and 200°F (82°C and 93°C) to ensure that the wax stays liquid, but not too hot.

3. Place the first napkin in the embroidery hoop, centering the area where the wax resist will go in the frame. The embroidery hoop is used to keep the fabric tight and to raise it up from the table so that the wax can go all the way through the fabric. Keep in mind that if the hoop is smaller than the intended design, you will need to move the hoop to complete the design. I would suggest a pattern that will work with the size embroidery hoop you are using. Trapping a waxed area between the hoop parts will probably leave a mark.

4. Now for the design, I like to have a concept in mind, but I tend to leave the actual details of the design to the moment. I love a bit of serendipity in my designs. For this project, I have chosen to do a repeating pattern with my rosette mold. Move the napkin as close to the hot wax as possible to prevent the stamp from cooling enroute to the napkin. Cool wax doesn't penetrate the fabric as well as hot wax. Also, work quickly. The longer you take to line up the stamp, the more likely it is to drip where it doesn't belong. Let the stamp warm up in the wax for a minute or two. Especially with napkins, which tend to be a slightly thicker fabric, it is important that the wax be hot and plentiful. Practice on a scrap piece of fabric before starting on the actual napkins.

NOTE: I like to use a large flour sack towel as a sample piece to try out different ideas, either stamps, or patterns or brush techniques. Once the fabric is filled up, I put the whole thing in a dye bath to see what the finished effect will be.

5. Once all the napkins have been printed, allow them to cool completely and prepare the dye bath according to package instructions. Submerge the napkins in cold water to get them completely wet and place all the napkins into the dye bath at once so that the color will be the same for all of them. Also throw in a couple sample scraps of fabric to use to check the color periodically. Once the color is the right density, remove the napkins from the dye bath and rinse them in cold water. Keep rinsing until the water runs clear.

6. Gently squeeze out the water and hang the napkins, allowing them to dry completely.

7. Once dry, iron or boil the napkins according to the instructions on page 105 to remove the wax. When deciding which method to use, consider the color fastness of the dye. Fragile dyes are probably best ironed to remove the wax.

BATIK DISH TOWELS

For this project, I decided to use a free-form technique that works well for brushes or tjanting tools. As with the napkins, I will be using 60 percent beeswax, 40 percent paraffin wax blend. How much is needed will depend on the size of the electric skillet and the tool used to apply the wax. If a brush will be used, the wax level should probably start out at about a ½" (1.3 cm) deep. To use a tjanting tool, the wax level needs to be deeper in order to be able to fill the tool with wax.

MATERIALS

flour sack towels

beeswax

paraffin wax

small electric skillet

cheap bristle brush or tjanting tool

cold water dye

dye vat (I like to use a stainless steel stock pot, but a plastic bucket should work as well)

clothesline and clothespins

embroidery hoop, large enough to hold part of the towel

iron and ironing board, optional

stainless steel stock pot, optional

1. Follow steps 1 through 3 of Batik Napkins on page 106 to prepare the dish towel, melt the wax, and place the dish towel in the embroidery hoop.

2. For this project, I have chosen to do a wavy lines motif. It is easy to do with minimal tools and doesn't require a lot of artistry. I do suggest that you place a sketch of the motif next to the towel to use as a reference while brushing on the wax. For my wavy lines project I didn't need this and used freeform lines.

3. There are a few things to keep in mind before actually putting wax to fabric.
• First, the wax will bleed a bit, so choose a brush size that is slightly smaller than the desired stroke width.
• Second, a brush does not hold much wax, and the start of the brush stroke will probably be fatter than the end. This can be used to an advantage, but needs to be taken into consideration when planning the start and stop of the line. This is less of an issue with a tjanting tool, since it has a built-in reservoir for wax.
• Third, the wax needs to be hot enough, probably between 180°F and 200°F (82°C and 93°C) to create a wax layer that goes all the way through the fabric. As soon as the brush is removed from the electric skillet, the wax will start to cool and at some point the wax will no longer penetrate the fabric. Ideally, this happens after the wax on the brush runs out.

Place the brush in the hot wax for about 15–30 seconds, or until the bristles heat up and find their natural shape. Paint the wax onto the towel, moving the hoop if needed. Once the design is completed, let the wax cool and prepare the dye bath.

4. Wet the waxed fabric with cold water first so that it will absorb the dye more evenly. Transfer the wet towel to the dye bath and soak it until the desired tint has been achieved. Then remove the towel from the dye bath and rinse with cold water until the water runs clear.

5. Hang the towel until it is completely dry. Once dry, it can be ironed or boiled to remove the wax depending on the preferred method and color fastness of the dye (see page 105 for details on wax removal).

ENCAUSTIC

Encaustic comes from the Greek word *enkaustikos*, which means "to burn in." The concept has been around for millions of years and it involves painting with hot beeswax, which may or may not have color added to it. Today, artists are doing more than just painting with it; beeswax is being used for collage and mixed media art pieces our forefathers never dreamed of.

What I love about encaustic is its diversity of uses. Encaustic can be used to create bold pieces with saturated colors; create light, ethereal works; or spare and monochromatic pieces that can all be equally stunning and highly individual. Pieces can be very dimensional or completely smooth. Encaustic can also be used to create mixed media pieces utilizing found objects to create one-of-a-kind works of art.

ENCAUSTIC MEDIUM

Although one can buy encaustic medium and encaustic paint ready-to-use, I have included the basics on how to make it here. Encaustic is essentially just beeswax and resin, but pigment may be added as well. The resin increases the melting temperature and adds luster, hardness, and shine once the piece is fully cured. The added resin also makes the paint more brittle, which is why I recommend using a nonflexible substrate, such as wood, tile, or a specialized encaustic board, for this project.

Although it is possible to use traditional yellow cappings wax, I strongly urge the use of "white" beeswax. When I first tried encaustics, I used yellow beeswax, which was what I had on hand. I added my pigments and thought I could work with the yellow undertones, but all I ended up with was muddy colors and very childish-looking artwork. It was a waste of wax and pigment! I now use only white, bleached beeswax and have had a much easier time creating works that I am proud of.

MASTER BATCH

There is artistry even in the making of the medium. The desired finished effect will determine the exact ratios of beeswax to resin. But keep in mind that if too much resin is used, the medium will become too brittle and could flake off. If not enough resin is used, the piece will be a bit soft and prone to dust accumulation. I like the proportion of 9 parts beeswax to 2 parts damar resin. Others prefer a ratio of 10:1.

MATERIALS

1 lb (450 g) "white" beeswax
100 g damar resin
small electric skillet with temperature control
spoon or stir stick
bread loaf pan with nonstick coating
large knife

1. Melt the beeswax in the electric skillet, heated to the 180°F–200°F (82°C–93°C) range. Once the wax is melted, add the resin and stir until all the resin is melted and incorporated. This will take a half hour or more, so be patient. There will be some impurities in the resin that will not melt. Ignore them for now.

2. Pour the wax and resin mixture into the loaf pan and allow the wax mixture to solidify. Cover the work surface with a layer of wax paper. Once the wax is hard enough to remove from the loaf mold, but still slightly warm to the touch, invert it over the wax paper.

3. The impurities that were in the resin sunk to the bottom and are now visible. Use the large knife to cut or scrape them off. I like to keep the wax scrapings and add them to the next batch of encaustic medium that I make so nothing goes to waste.

NOTE: Damar resin can be purchased at art supply stores either locally or online.

4. Use the knife to cut the clean portion of the beeswax/resin block into smaller cubes that are easier to use as is, or are the perfect size to toss into a tin and mix with color.

MIXING ENCAUSTIC PAINT

In many ways, encaustic paints are like watercolor and oil paints. When there is just a tiny bit of pigment used, the end result is more of a wash, as with watercolors. If a lot of pigment is used, the end result is more like oil paints, creating super-saturated vibrant colors. I use French mineral pigments, which have some transparency and a nice range of colors, but any kind of powdered artists' pigment will work.

MATERIALS

beeswax/resin blocks

powdered artists' pigment in colors of your choice

several 4 oz (113 g) flat-bottomed tins

mini loaf pan

electric griddle with temperature control

paint brush

1. Heat the electric griddle to 200°F (93°C). Make sure the temperatures does not rise above 220°F (104°C), or the wax may begin to smoke and degrade.

2. Since I make my own medium and I have already cut the beeswax/resin blocks into manageable pieces, I just toss a couple of chunks into a mini loaf pan and put the pan on my hot griddle.

3. As the beeswax/resin blocks melt, I add a touch of artists' pigment to a 4 oz (113 g) flat-bottomed tin.

NOTE: If I am making saturated, super-strength colors, I like to mix them in 4 oz (113 g) flat-bottomed tins, either blending them with encaustic medium directly on the electric griddle or adding the concentrated color to a mini loaf pan and diluting it as desired.

4. Then I add pigment to the melted beeswax/resin and stir with my brush until it is fully mixed. I keep a piece of absorbent paper on hand to test the color density. It is much easier to add more pigment than it is to increase the size of the color batch by adding additional encaustic medium. Now you're ready to paint!

PAINTING WITH ENCAUSTICS

This is an easy beginner project that utilizes the paints you just mixed. Since this is artwork, there are no rules on what can or cannot be done, but there are a few guidelines.

MATERIALS

sized panel

encaustic paints in tins

natural bristle brushes (I like using 1–2" [2.5–5 cm] brushes for overall coverage and smaller brushes for more intricate work)

butane torch or heat gun

Joss papers or silver leaf

burnishing tool

paper towel or nylon stocking

electric skillet with temperature control

Kraft paper

After applying each layer of wax, it must be fused to the layer below it using either a heat gun or a small butane torch. I prefer the latter for most of my work. Fusing the wax is as easy as quickly moving the torch over the wax and watching it go from matte to shiny. That's all the heat that is needed; any more, and the wax will move or develop heat spots. It takes a bit of practice to master the amount of heat and the methodology, but it is a fairly easy learning curve.

Observe proper safety precautions. Make sure the space is properly ventilated. Wear gloves to protect hands from hot wax and concentrated pigments. Never leave the electric griddle or any other electrical tools plugged in and unattended. Never let the encaustic medium and paint exceed 220°F (104°C). Keep a fire extinguisher on hand, just in case.

Make sure that the encaustic medium or paint has turned matte before adding another layer. A matte finish means it has cooled enough to apply more wax.

1. For this project we will put down lots of translucent layers to give the piece depth. To start, pick a color palette of three colors. I chose turquoise green, lavender blue, and blackcurrant red and I mixed just the amount needed for this project in my mini loaf pans, adding just a touch of color and painting it on my sample Kraft paper to make sure it was what I wanted—a light wash.

2. Warm the panel with the heat gun to make it easier to apply the wax. Brush the encaustic paint onto the panel in the three colors selected, covering the whole panel from left to right in random stripes.(A) Fuse this layer with the heat gun or torch. (B)

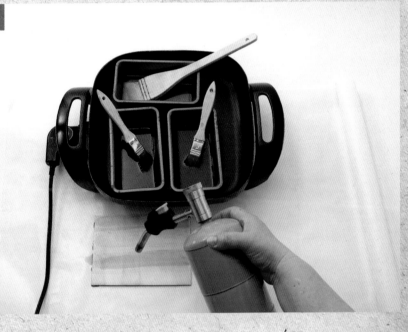

3. Rotate the piece clockwise 90 degrees and paint similar stripes going the other direction. Fuse to the layer below.

4. Apply multiple layers on top of the first layers, always turning the panel clockwise 90 degrees and fusing with the heat gun before adding the next layer. Since the colors are somewhat translucent, they can be layered with great effect. Don't be afraid to experiment. Notice that with each layer the color gets darker and a bit more saturated.

5. Once all the layers have been painted on and fused, I added a small square of silver to float on top. It is a fun, visually interesting touch that you can add using silver leaf, but I have chosen to use Joss papers. Joss papers, which are made to be burnt as offerings in various Asian religious ceremonies, can be found at Asian grocery stores or online. Ignore the fact that some are silver colored and some are gold. All will yield a silver result.

6. To do the transfer using the Joss paper, cut out a square of silver to the desired size. If the panel is still very hot, allow it to cool a bit, but if it has been sitting longer than 30 minutes, use the torch to warm the spot where the silver will go. Be careful to just warm the wax, making it slightly sticky, not so soft that it will move around when applying the silver. If using silver foil, no additional sizing is needed, just apply according to directions.

7. Place the Joss paper piece face down on the artwork and rub with a burnishing tool. Once the whole square has been burnished, remove the backing paper by adding a couple drops of water and rubbing the paper gently with your fingertips. It will start to ball up and lift off. Continue until all the paper has been removed.

8. Make one more pass over the silver square with the torch to remove any residual paper fibers that could cloud the piece later.

9. Set the piece aside to cool and cure completely for a couple days and then buff it with a paper towel or nylon stocking. Stand back and admire!

ENCAUSTIC COLLAGE

For this project, I use a collection of printed paper napkins and uncolored encaustic medium. There are some wonderful graphic napkins out there that can easily be turned into works of art. Look for napkins with a good proportion of white space and great graphic elements.

MATERIALS

- encaustic panel
- napkins in a variety of patterns
- other natural elements such as leaves, thread, wire mesh
- encaustic medium
- natural bristle brush
- electric griddle
- butane torch or heat gun
- nylon stocking or paper towel
- Kraft paper

1. To prepare the napkins for use, first remove the backing paper. Most commercially available napkins have one or two layers of white paper attached to the graphic paper on the front. Remove those extra white sheets. An easy way to do this is to take a piece of painter's tape or masking tape and stick it to the white backing paper and pull. The paper will rip, but that's okay, since it will not be part of this project. If there is another piece of white paper, remove it the same way. Now, all that's left is the super thin, graphic paper that can be added to encaustic panels.

2. In order for the paper to stick, the panel needs to be sized with several layers of encaustic medium. Warm the encaustic medium and the brush that will be used for applying the medium in a mini loaf pan on the electric griddle.

3. Before applying the wax, warm the panel with the butane torch or heat gun. Then cover the panel with medium using brush strokes that go from one side of the panel to the other until the entire panel is covered. Use the torch to fuse the medium to the board. Turn the piece clockwise 90 degrees and repeat, adding another layer and fusing with the torch. Keep doing this until four or five layers are built up. The piece should be relatively smooth. (See photos at right.)

4. With napkin pieces torn or cut into interesting shapes to capture all or part of a graphic element, begin to plan the arrangement of the pieces on the board. Keep in mind that the white areas will be almost transparent when medium is applied, so adding just a couple elements with each layer works well. Then, as the layers progress, the overlap of papers becomes really interesting. Elements that got added early on will start to recede into the background as additional layers of wax are added to the top. This can be used to great effect.

5. Toward the end, consider adding other elements to the collage. Most, if not all, of the elements need to be embedded in wax to hold them in place. Take care when fusing around protruding elements that are not completely waxed, such as dried leaves, since they might burn. It is best to wax the element and then apply at least part of it to the art board.

6. Once the work is complete, let it cure for a couple days and then buff it with a nylon stocking or paper towel.

CHAPTER 8
INGREDIENTS GUIDE

BUTTERS

COCOA BUTTER

INCI: Theobroma Cacao (Cocoa) Seed Butter

This is a hard butter with a melting temperature of 100°F (38°C). It has a relatively long shelf life of two to five years. It also contains polyphenols and phyosterols and some skin softening Vitamin E. Cocoa butter provides an occlusive layer on our skin.

Palmitic Acid 25–30%

Stearic Acid 31–35%

Oleic Acid 34–36%

Linoleic Acid 3%

MANGO BUTTER

INCI: Mangifera Indica (Mango) Seed Butter

This is a medium-hard butter with a melting temperature of 86°F–98.6°F (30°C–37°C) and a slight sweet scent. It has a shelf life of two to three years. It has emollient properties, high oxidative ability, wound healing, and regenerative activity.

Palmitic Acid 25-30%

Stearic Acid 31–35%

Oleic Acid 34–36%

Linoleic Acid 3%

SHEA BUTTER

INCI: Butyrospermum parkii (Shea Butter) Fruit

This is a soft butter with a melting temperature of 97°F–100°F (36°C–38°C). It has a shelf life of two to three years. Shea penetrates deep into the skin, rejuvenating damaged cells and restoring elasticity and tone. In soap, it has a larger than average proportion of fats that are not saponifiable, meaning the soap will actually retain some moisturizing qualities.

Palmitic Acid 25–30%

Stearic Acid 31–35%

Oleic Acid 34–36%

Linoleic Acid 3%

OILS

*APRICOT OIL

INCI: Apricot Kernel (Prunus Armeniaca) Oil

Very similar to sweet almond oil, apricot oil is a lightweight oil with a shelf life of about one year. On skin it is softening, moisturizing, and is easily absorbed. In soap it creates a conditioning stable lather. It is often used in place of sweet almond oil for those with nut allergies.

Palmitic Acid 4–7%	
Stearic Acid 1%	
Oleic Acid 58–78%	
Linoleic Acid 20–34%	
SAP VALUE KOH / NaOH: .190–.135	

*AVOCADO OIL

INCI: Persea Gratissima (Avocado) Oil

Medium-weight oil with a shelf life of about one year. Avocado oil is easily absorbed into deep tissue, and with its wonderfully emollient properties, it's ideal for mature skins. It also helps to relieve the dryness and itching of psoriasis and eczema, and it helps with skin regeneration.

Palmitic Acid 10%	
Stearic Acid 4%	
Oleic Acid 75–80%	
Linoleic Acid 7–10%	
SAP VALUE KOH / NaOH: .190–.135	

CASTOR OIL

INCI: Ricinus Communis (Castor) Seed Oil

One of the thickest and heaviest oils available, castor oil has a shelf life of five years. On skin it is a humectant and helps to retain moisture levels. In soap it makes for a softer bar but improves lather when used in small quantities. Can also make the soap a bit translucent.

Oleic Acid 2–6%	
Linoleic Acid 2–4%	
Ricinoleic Acid 85%	
SAP VALUE KOH / NaOH: .180–.128	

*Not used in recipes in book, but useful substitutions

COCONUT OIL

INCI: Cocos Nucifera (Coconut) Oil

Solid at temperatures below 76°F (24°C), this is a medium-weight oil with a shelf life of about 2–4 years. On skin, coconut oil contains some very potent antioxidants that can help with skin aging, and help repair damage caused by light and radiation. It softens, moisturizes, and soothes chapped and itchy skin. In soap, coconut oil creates loads of big bubbles. Although it makes a hard bar, it melts very quickly in water, making it a great soap for hard water conditions. It is also extremely cleansing and in high proportion, can be very drying to the skin.

Lauric Acid 48%	
Myristic Acid 18%	
Palmitic Acid 9%	
SAP VALUE KOH / NaOH: .2660 / .181	

EMU OIL

INCI: Emu Oil

This medium-weight oil has a shelf life of one year. Emu oil has wonderful skin-softening and cell regenerating properties and is great for restoring the skin's barrier function. This oil also helps other ingredients absorb better into the skin. Although it can be used in soap, I recommend that this oil be reserved for direct applications.

Palmitic Acid 25–30%	
Stearic Acid 31–35%	
Oleic Acid 34–36%	
Linoleic Acid 3%	
SAP VALUE KOH / NaOH: .135 / .192	

*FRACTIONATED COCONUT OIL

INCI: Caprylic/Capric Triglyceride

This is an incredibly lightweight oil with at least a 2-year shelf life. Since this is a portion of coconut oil on a molecular level (fractionated means that a number of long chain fatty acids were removed chemically), the resultant moisturizing oil is easily absorbed by skin and hair.

Caprylic Acid 53–55%	
Capric Acid 36–47%	
SAP VALUE KOH / NaOH: .334 / .237	

GRAPE SEED OIL

INCI: Vitis vinifera (Grape) Seed Oil

This is another great lightweight oil. It has a relatively short shelf life of 6 months to 1 year. On skin it can help to retain moisture and reduce inflammation and itchiness. I recommend that it is used sparingly in soap, since it has a short shelf life.

Palmitic Acid 7%	
Stearic Acid 4%	
Oleic Acid 16%	
Linoleic Acid 72%	
SAP VALUE KOH / NaOH: .187 / .133	

JOJOBA OIL

INCI: Simmondsia Chinensis (Jojoba) Seed Oil

Technically this is not an oil, but rather a liquid wax ester. It is lightweight and has a long shelf life of at least 2 years. It is great for hair and skin, since it mixes with sebum and allows it to be washed away. It is an excellent emollient. In soap, jojoba adds moisturizing and conditioning properties as well as extending the shelf life of the soap.

Palmitoleic Acid 2%	
Stearic Acid %	
Oleic Acid 14–25%	
Gadoleic Acid 37%	
Erucic Acid 20%	
SAP VALUE KOH / NaOH: .065 / .091	

OLIVE OIL

INCI: Olea Europaea (Olive) Fruit Oil

This a medium-weight oil with a shelf life of about one year. It is known for its skin moisturizing and nourishing properties. It makes an extremely gentle soap, suitable for sensitive skin types. Used by itself in soap, it yields a lather that is creamy rather than bubbly and while still fresh, the lather may actually feel a bit slimy. That sliminess goes away after a 6-month or more cure time.

Palmitic Acid 10%	
Stearic Acid 3%	
Oleic Acid 55–83%	
Linoleic Acid 4–21%	
Linolenic Acid 1%	
SAP VALUE KOH / NaOH: .135 / .190	

*RICE BRAN OIL

INCI: Oryza Sativa (Rice) Bran Oil

Rice bran oil is a medium-weight oil with a shelf life of 6 months to one year. It is high in Vitamin E and fatty acids that make skin soft and improve elasticity. It also helps with cell regeneration. In soap, it behaves similarly to olive oil, but I find it adds silkiness to the soap and makes a slightly harder, longer lasting bar.

Palmitic Acid 16%	
Stearic Acid 2%	
Oleic Acid 42%	
Linoleic Acid 36%	
Linolenic Acid 2%	
SAP VALUE KOH / NaOH: .135 / .190	

SAFFLOWER OIL

INCI: Carthamus Tinctorius (Safflower) Seed Oil

This is a lightweight oil that has a 2 year shelf life. It is a nourishing oil that is great for oily skin and is easily absorbed. In soap, it is mild and moisturizing.

Palmitic Acid 3–6%	
Stearic Acid 1–4 %	
Oleic Acid 13–21%	
Linoleic Acid 73–79%	
SAP VALUE KOH / NaOH: .135 / .190	

SESAME OIL

INCI: Sesamum Indicum (Sesame) Seed Oil

Sesame seed oil is a lighter-weight oil with a about one year shelf life. It is frequently used in massage products, since it is doesn't stain clothes. It is also good at moisture regulation, reducing inflammation, and cell regeneration. In soap, sesame oil will be moisturizing and conditioning.

Palmitic Acid 9%	
Stearic Acid 4%	
Oleic Acid 45%	
Linoleic Acid 40%	
SAP VALUE KOH / NaOH: .135 / .191	

SOYBEAN OIL

INCI: Glycine Soja (Soybean) Oil

Soybean oil is a lightweight oil with about a one year shelf life. It is easily absorbed and can soothe chapped skin. In soap it is mild, moisturizing, and gives a creamy lather. Soybean oil is also readily available and economical. I prefer to use organic soybean oil, especially in my direct skin applications.

Palmitic Acid 10%	
Oleic Acid 29%	
Linoleic Acid 53%	
Linolenic Acid 8%	
SAP VALUE KOH / NaOH: .134 / .188	

SUNFLOWER OIL (HIGH OLEIC)

INCI: Helianthus Annuus (Sunflower) Seed Oil

High oleic sunflower oil is a medium-weight oil with a shelf life of about one year. On skin, it has great softening and anti-inflammatory properties. In soap, it works nicely with palm and olive oils to give a rich, creamy lather that's very moisturizing.

Palmitic Acid 5–7%	
Stearic Acid 3–6%	
Oleic Acid 15–36%	
Linoleic Acid 61–73%	
SAP VALUE KOH / NaOH: .136 / .191	

SWEET ALMOND OIL

INCI: Prunus dulcis (Almond) oil

This is a lightweight oil with a shelf life of one year. This emollient oil softens and nourishes the skin, and is often used in massage. In soap it creates a conditioning, stabile lather.

Palmitic Acid 2–6%	
Palmitoleic Acid 2%	
Stearic Acid 3%	
Oleic Acid 60–78%	
Linoleic Acid 10–30%	
Linolenic Acid 2%	
SAP VALUE KOH / NaOH: .137 / .194	

*Not used in recipes in book, but useful substitutions

MISCELLANEOUS

E-WAX NF

INCI: Cetearyl Alcohol (and) Polysorbate 6

E-wax is an all-in-one, easy-to-use emulsifier used to combine water and oil into a lotion or cream. NF means that it conforms to the National Formulary guidelines.

LANOLIN

INCI: Lanolin

Technically, lanolin is a wax. It is the yellow, extremely viscous substance sheep produce that helps them shed water from their coats. Most lanolin is a by-product of wool processing, since it needs to be removed before the wool can be dyed or woven. For skin, it is a wonderful protectant, creating a semi-permeable layer to keep the elements out and the moisture in.

LECITHIN

INCI: Lecithin

Derived from soybeans, lecithin is usually added to act as an emulsifier, but it also helps to replenish and repair the skin. It works wonders on nails.

OPTIPHEN

INCI: Phenoxyethanol (and) Caprylyl Glycol

Optiphen is a broad-spectrum, paraben-free preservative that can be used to preserve emulsified products that contain water, such as lotions and creams. There are other preservatives, but this is one of my favorites.

ROSEMARY OLEORESIN

INCI: Rosmarinus Officinalis (Rosemary) Leaf Extract

Rosemary oleoresin, or ROE as it is often abbreviated, is added to oils as an antioxidant. It helps prevent oils from oxidizing and becoming rancid and it prolongs the shelf life of a product. This is not a preservative.

STEARIC ACID

INCI: Stearic Acid

Stearic acid is one of many fatty acids that occur naturally in the plant and animal world. It is a component in animal tallow, cocoa butter, and vegetable fats. When it's used in cosmetic products, stearic acid primarily fulfills the role of a thickener or hardener. Stearic acid is the substance that helps a bar of soap retain its shape.

T-50 (VITAMIN E)

INCI: Tocopherol

T-50 is an all-natural, blend of mixed tocopherols and edible vegetable oils.

It functions as an excellent antioxidant for oils or products that contain oil. It helps delay oxidation and will help to extend the shelf life of oils and oil-based products. It works a bit differently than the rosemary oleoresin, so I often use both.

FATTY ACID GLOSSARY

ERUCIC ACID (C22:1):

Erucic acid is used in cosmetic products as an emollient because it provides a protective layer for skin.

GADOLEIC ACID (C20:1):

Gadoleic Acid prevents transdermal water loss. It serves as an occlusive barrier without being sticky or overly greasy.

LINOLEIC ACID (C18:2):

Linoleic acid helps to improve skin's barrier function, helps to soothe itchy dry skin, and acts as an anti-inflammatory and a moisture retainer.

OLEIC ACID (C18:1):

Very moisturizing, oleic acid helps skin cells regenerate quickly. It is very easily absorbed by the skin and it acts as an anti-inflammatory.

PALMITIC ACID (C16):

Palmitic acid helps to reinforce skin barrier function, and is a very good emollient.

PALMITOLEIC ACID (C16:1):

A building block in our skin that prevents burns, wounds, and scratches. Most active microbial in human sebum, palmitic acid treats damaged skin and mucous membranes.

RICINOLEIC ACID (C18:3, N-9):

Only found in castor oil, it has analgesic and anti-inflammatory effects.

STEARIC ACID (C18):

Stearic acid helps improve moisture retention; increase flexibility of the skin, and repair skin damage.

VITAMIN E:

This vital vitamin sinks into our skin and behaves as an antioxidant.

HERBS

ARNICA
(ARNICA MONTANA)

Arnica is known for its natural anti-inflammatory properties and its ability to reduce swelling. Arnica salve is a muscle workhorse. It is wonderful for sprains, strains, and pulled ligaments or muscles! It also works really well to help bruised areas heal.

CALENDULA BLOSSOMS
(CALENDULA OFFICINALIS)

Calendula is anti-inflammatory and anti-microbial; good for minor cuts and burns. Promotes healing by aiding clotting.

CHAMOMILE, GERMAN
(MATRICARIA CHAMOMILLA)

Contains the flavonoids apigenin, luteolin, and quercetin, and the volatile oils alpha-bisabolol and matricin, which help to make chamomiles anti-inflammatory, antispasmodic, and antioxidant. It is also known for being calming, soothing, and healing.

CHAMOMILE, ROMAN
(ANTHEMIS NOBILIS)

See German Chamomile, which has a very similar effect on skin.

CHICKWEED LEAF
(STELLARIA MEDIA)

Great soother of itchy and sore skin. It helps to cools and relieve inflamed areas.

COMFREY LEAF AND ROOT
(SYMPHYTUM OFFICIALIS)

The high allantoin content in comfrey is the star of this herb. It encourages the renewal of skin cells and the strengthening of skin tissues, and it helps to heal wounds as well as to reduce the visible effects of aging by bringing back elasticity to the skin. It is also a mild astringent, which makes comfrey great for sunburn, ulcers, and sores. Since it is so powerful, it should not to be used on cuts that are deep, since it can trap infection by healing the top layer of skin faster than deeper layers.

LAVENDER
(LAVANDULA ANGUSTIFOLIA)

Lavender is a super star for burns and encouraging wound healing. It has also been found to be effective against the principal bacteria involved in acne and is known to soothe insect bites and bee stings.

MARSHMALLOW ROOT
(ALTHAEA OFFICINALIS)

Soothes skin and promotes wound healing.

PLANTAIN LEAF
(PLANTAGO MAJOR)

Astringent, antibacterial, anti-inflammatory, and anti-itch, plantain has a long history of being used for problem skin. It contains salicylic acid, which helps to reduce acne breakouts. Plantain is also great for bee stings. Just pick a leaf, mash it up a bit, and apply as a poultice to the sting.

ROSEMARY
(ROSMARINUS OFFICINALIS)

Removes excess oil from skin without causing excess dryness. It has regenerating and stimulating effects that help to rejuvenate skin and give it a more youthful glow. It helps to restore elasticity and firmness to the skin and encourages skin cell turnover and renewal.

SELF-HEAL
(PRUNELLA VULGARIS)

With astringent, antibacterial, and antiseptic properties, self-heal helps to reduce pain and promote healing of cuts and scrapes. Can even be used on skin rashes.

ST. JOHN'S WORT
(HYPERICUM PERFORATUM)

Helps to support skin elasticity, soothes redness and irritation, and works wonders on burns, wounds, insect bites, and other skin irritations.

THYME
(THYMUS VULGARIS)

Thyme's natural antimicrobial, antibacterial, and anti-inflammatory properties help to remove excess oil and dirt from pores and reduce acne outbreaks. Thyme contains powerful antioxidants that can help protect skin from premature aging.

ESSENTIAL OILS

(T=top note H=heart note B=base note)

BENZOIN (B)

It has a sweet, warm, vanillalike odor that is long lasting and makes it an excellent fixative.

BERGAMOT (T, M)

Cold-pressed essential oil produced by the rind of a bergamot orange fruit. Bergamot essential oil assists with stress, tension, SAD, PMS, skin problems, anxiety, and colds and flu. On the skin, Bergamot oil has antiseptic and astringent properties, which make it useful after shaving.

BERGAMOT MINT (T)

Sweet, herbaceous, and slightly floral, it offers a smooth mint scent, tempered by floral, fruity, notes. The scent is known to be calming and relaxing.

BIRCH (H)

Sweet, sharp, minty, fresh, camphoraceous aroma, it is distilled from the bark. It has powerful pain relieving, anti-inflammatory, and antispasmodic properties, which makes it great for use on the joints and muscles. Exercise caution when using it though, as birch oil can sometimes irritate the skin. It should always be used carefully and in the correct dilution.

CAMPHOR (T)

An excellent oil for insect control and pain management. Its high camphor content also makes a great clearing oil in skin care.

CEDARWOOD, ATLAS (H)

Cedarwood is steam distilled from the wood. It is soft, smooth, and grounding. and uplifting, yet calming. The fragrance is delicate and sweet with a woody undertone. It is known for its detoxing uses, where it offers a gentle stimulating effect, and the breakdown and elimination of toxins, including fatty build up, as in cellulite. It is also effective on sore muscles and joints, and it is an upper and lower respiratory decongestant.

CHAMOMILE, GERMAN (H)

On the skin, it is a miracle worker and calms red, dry, and irritated skin, as well as calming skin allergies, such as eczema, psoriasis and all other flaky skin problems. It is also a great tissue regenerator. Chamomile has the highest content of chamazulene, which makes it very anti-inflammatory. Its scent has a warm, somewhat sweet, herbaceous note

that blends well with most flower oils, most of the citrus oils, and all of the other herbal oils.

CLARY SAGE (B)

Clary sage essential oil assists with muscle pains, anxiety, stress, nervous tension, insomnia, menstrual problems, and PMS. On the skin, clary sage can help restore lost skin tone. The scent is fresh and herbaceous, with grassy notes that dry down into amber notes, due to high sclareol compounds.

CORIANDER (T)

Coriander essential oil can help assist the digestive system, ease rheumatism and arthritis pain and muscular spasm, while detoxifying the body.

CYPRESS (H)

This is a grounding, centering, essential oil, well-known for helping one to get emotional balance and mental clarity. On skin it can help with sluggish lymphatic conditions and circulatory treatments to eliminate toxins from the system. The scent is very evergreen with some herbaceous notes.

EUCALYPTUS (T)

This oil is an intensely clean, refreshing, and uplifting oil. It is commonly used to treat upper respiratory conditions. It is also a great choice for muscle pain.

FENNEL (T)

Fennel essential oil has a sweet, earthy, aniselike aroma. It is a good skin tonic and can be beneficial for dull, and tired skin. It can also help with puffy areas (i.e., eyes and thighs), as it helps to reduce water retention.

FRANKINCENSE (B)

This oil is used to rejuvenate skin and encourage skin cell growth. It is also effective in healing sores and wounds, and helps to reduce scarring and skin inflammation. Frankincense soothes and calms the mind. The scent is fresh and earthy, reminiscent of incense and has a slight sweetness that is both warm and deep. In perfumery it can be a powerful bridge note to tie the components together and prevent individual notes from standing out.

GRAPEFRUIT, PINK (T)

This oil has the ability to flush fluids out of tissues, making it a great choice for cellulite and fluid retention problems. It is also a great choice for congested, oily, skin where it can help balance and cleanse the skin without causing the skin to overreact and create more oil. Its scent is fresh and tart, with smooth citrus notes.

HELICHRYSUM (B)

This oil is known for it is superior antihematoma and scar healing benefits. It is a powerful circulation enhancer and skin regenerator, and it is great for the treatment of bruises and poor circulation. The scent is warm, herbaceous, and slightly sweet.

JUNIPER BERRY (T)

This oil has diuretic effects that help alleviate fluid retention. It has disinfectant and purifying actions and is often combined with white sage for spiritual clearing and cleansing. It also improves circulation and can help with muscle aches, since it helps eliminate lactic acid buildup at the heart of pain and inflammation. The scent is fresh and uplifting with woody notes.

LAVANDIN (H)

A hybrid from true lavender and spike lavender. Very similar to lavender, but smell is a bit more camphorus.

LAVENDER (H)

As an essential oil it soothes the spirit, relieves insomnia, and relieves muscular spasms. Lavender Absolute stimulates the growth of new skin cells and tones the skin. It is a must have in the first aid kit and is one of the few oils that can be applied "neat" (directly to the skin without dilution).

LEMON (T)

This oil has an uplifting and refreshing effect as well as relieving some anxiety and stress. On skin, it helps with cleansing and keeping oily skin under control. The scent is bright and fresh.

MENTHOL CRYSTALS (T)

Menthol crystals are an organic compound obtained from peppermint or other mint oils. It is solid at room temperature and melts slightly above. Menthol is used for its local anesthetic and counterirritant qualities.

MYRRH (B)

This oil often combined with Frankincense for meditation and reflection. On skin it is nurturing to mature and dry skin, and offers some tightening and smoothing properties. The scent is deep, earthy, and slightly balsamic.

NUTMEG (T)

This oil has a sharp, spicy, and rather musky aroma. It has been shown to be a good anti-inflammatory that is successful in relieving pain, especially muscular aches and pain, as well as rheumatism.

ORANGE, SWEET (T)

This oil's gentle, clarifying nature cheers the heart and brightens the mood. It is a great choice for oily and congested skin where it can help remove excess oil without over drying. It is also a popular oil for cleaning products.

PALMAROSA (B)

This oil is well-known for its use as an antiviral, due to it is high geraniol content. Geraniol is one of the most powerful of nature's antivirals. On skin, it is great for sensitive and antiaging blends. The scent is sweet, warm, slightly green, and bit roselike. It is a nice addition to any floral blends.

PATCHOULI, DARK (B)

This oil has a very calming effect on emotions. Works well on dry skin and hair. In scent blends it helps give the blend some staying power.

PEPPERMINT (T)

This oil has a soothing effect on the nervous system by relieving tension and irritation. It works well as a decongestant, helping to open up completely blocked sinuses. Great headache reliever.

PINE NEEDLE (H)

This oil has antimicrobial and decongesting benefits. It also helps relieve aches and pains associated with muscle tension. The scent is woody, but with depth and sweetness.

ROSE ABSOLUTE (H)

This absolute has excellent emollient and hydrating properties, helping the skin stay soft and properly moisturized. It is also a stimulating oil, which is great for fighting aging and maintaining a soft, dewy, and youthful complexion.

ROSEMARY (T)

This oil is great for detoxing, energy, as well as any treatment products geared toward pain relief. The scent is herbaceous and woody and blends well with other conifers and the mints.

SANDALWOOD (B)

This oil is known for its supportive effect on tissue, being effective for dry, chapped, cracked skin, rashes, and acne. The scent is sweet and woody with warm and rich notes. It also has a definite musky note.

SPRUCE (T)

This oil has detoxing capabilities, working to eliminate lactic acid buildup in tissue, which is at the heart of aching, sore, muscles. The scent has a coniferous note and is clean, sweet, and slightly green.

TANGERINE (T)

This oil has an uplifting and refreshing effect as well as relieving some anxiety and stress. On skin, it helps with cleansing and keeping oily skin under control. The scent is bright, fresh, and sweet, and gives blends a lift.

TEA TREE (H)

This oil is another one for the first aid kit. It is one of the most active antiseptic oils available. It soothes insect bites, calms inflammation, acts as an immune system stimulant, and is an effective broad antimicrobial agent against bacteria, mold, and fungus.

VANILLA CO2 (B)

This is the carbon dioxide extract of the vanilla bean. Vanilla has few actual therapeutic benefits, but is important for its fragrance. It is calming, soothing, and it works well in romantic and aphrodisiac blends. The scent is smooth, sweet, and exotic.

VETIVER (B)

This oil is well-known for having a calming effect on the psyche. It is a great choice for any blends geared toward mental health or depression. On skin it is great for mature, damaged, and dry skin. The scent is deep and earthy with just a touch of sweetness. In perfumery it is a powerful scent fixative.

YLANG YLANG (H)

This oil is known as a relaxation aid that reduces stress and tension. On the skin, ylang ylang oil has a soothing effect and it is often used to even out over-dry as well as overly oily skin by balancing the secretion of sebum. Its deep and exotic scent has a definite sweetness and smoothness, making it popular in perfumes.

RESOURCES

This is a list of suppliers I use in my own business and recommend. You will notice that there are many companies that are listed under multiple categories. I have chosen to list suppliers this way to help you easily find the ingredient for which you are looking. Many of the suppliers listed also carry items from other categories included here that may be very good, but since I have not used them, I cannot recommend them.

BEESWAX

BEEHIVE ALCHEMY
www.beehivealchemy.com
2115 N 56th St
Milwaukee, WI 53208
414-403-1993

DADANT & SONS
www.dadant.com
51 South 2nd
Hamilton, IL 62341
888-922-1293

LYE (SODIUM HYDROXIDE)

AAA CHEMICALS
www.aaa-chemicals.com
226 TEXAS CITY WYE
LaMarque, TX 77568
409-908-0070

BOYER CORPORATION
www.boyercorporation.com
P.O. Box 10
La Grange, IL 60525
800-323-3040

WHOLESALE SUPPLIES PLUS
www.wholesalesuppliesplus.com
10035 Broadview Rd.
Broadview Heights, OH 44147
800-359-0944

BASE OILS AND BUTTERS

COLUMBUS FOODS/SOAPERS CHOICE
www.soaperschoice.com
30 E. Oakton Street
Des Plaines, IL 60018
800-322-6457 Ext: 8930

BRAMBLE BERRY
www.brambleberry.com
2138 Humboldt St.,
Bellingham, WA 98225
360-734-8278

WHOLESALE SUPPLIES PLUS
www.wholesalesuppliesplus.com
10035 Broadview Rd.
Broadview Heights, OH 44147
800-359-0944

CAMDEN-GREY ESSENTIAL OILS, INC
www.camdengrey.com
3567 NW 82 Avenue
Doral, FL 33122
305-500-9630

NEW DIRECTIONS AROMATICS
www.newdirectionsaromatics.com
6781 Columbus Road
Mississauga, Ontario, L5T 2G9
Canada
800-246-7817

ESSENTIAL OILS AND FRAGRANCE OILS

AROMA HAVEN/RUSTIC ESCENTUALS, LLC
http://rusticescentuals.com
1050 Canaan Road
Roebuck, SC 29376
814-217-7113

NEW DIRECTIONS AROMATICS
www.newdirectionsaromatics.com
6781 Columbus Road
Mississauga, Ontario, L5T 2G9
Canada
1-800-246-7817

WHITE LOTUS AROMATICS
www.whitelotusaromatics.com
941 E. Snowline Dr.
Port Angeles, WA 98362
360-452-9516

CAMDEN-GREY ESSENTIAL OILS, INC
www.camdengrey.com
3567 NW 82 Avenue
Doral, FL 33122 US
305-500-9630

CONTAINERS AND PACKAGING

SKS BOTTLES & PACKAGING, INC
www.sks-bottle.com
2600 7th Avenue
Building 60 West
Watervliet, NY 12189
518-880-6980

WHOLESALE SUPPLIES PLUS
www.wholesalesuppliesplus.com
10035 Broadview Rd.
Broadview Heights, OH 44147
800-359-0944

BOTTLES AND MORE
www.bottlesandmore.com
4523 Sunset Oaks Dr.
Paradise, CA 95969
530-872-2745

HERBS AND BOTANICALS

MONTEREY BAY SPICE CO
www.herbco.com
241 Walker St.,
Watsonville, CA 95076
831-426-2808

ROCKY MOUNTAIN SPICE CO
www.rockymountainspice.com
3850 Nome St.,
Denver, CO 80239
303-308-8066

MOUNTAIN ROSE HERBS
www.mountainroseherbs.com
P.O. Box 50220
Eugene, OR 97405
800-879-3337

SPECIALTY COSMETIC INGREDIENTS

THE HERBARIE AT STONEY HILL FARM, INC.
www.theherbarie.com
Prosperity, SC 29127
803-364-9979

LOTIONCRAFTER
www.lotioncrafter.com
532 Point Lawrence Road
Olga, WA 98279
360-376-8008

INGREDIENTS TO DIE FOR
www.ingredientstodiefor.com
5103-A Commercial Park Dr.
Austin, TX 78724
512-535-2711

MAJESTIC MOUNTAIN SAGE
www.thesage.com
2490 South 1350 West
Nibley, UT 84321
435-755-0863

WHOLESALE SUPPLIES PLUS
www.wholesalesuppliesplus.com
10035 Broadview Rd.
Broadview Heights, OH 44147
800-359-0944

BRAMBLE BERRY
www.brambleberry.com
2138 Humboldt St.,
Bellingham, WA 98225
360-734-8278

INTERNATIONAL SUPPLIERS

VOYAGEUR SOAP AND CANDLE CO.
www.voyageursoapandcandle.com
#14 - 19257 Enterprise Way
Surrey, BC V3S 6J8
800-758-7773

GRACEFRUIT
www.gracefruit.com
146 Glasgow Road
Longcroft, Stirlingshire, FK4 1QL, UK
Telephone: 01324 841353

THE SOAP KITCHEN
www.thesoapkitchen.co.uk
The Soap Kitchen Plus (shop),
11a South Street,
Torrington, Devon,
EX38 8AA, UK
Tel: 01805 622221

SIDNEY ESSENTIAL OIL COMPANY
www.seoc.com.au
11 Burrows Road South,
St. Peters, NSW 2044
Australia

NEW DIRECTIONS AUSTRALIA
www.newdirections.com.au
47 Carrington Rd.
Marrickville, Sydney, NSW 2204
Australia

KOSMETISCHE ROHSTOFFE
http://www.kosmetische-rohstoffe.de
Kai Habermann
Am Grenzberg 34 63654 Büdingen
Germany

ACKNOWLEDGMENTS

I would like to thank my boyfriend Karl for discovering his passion for bees and for finding my stash of handmade soaps all those years ago. It was this combination of events that led to building Beehive Alchemy into the business it is today. I would also like to thank Martha Stewart, who's American Made project served as a jumping-off point for many of the opportunities that I have had the privilege to participate in, in the last year, the writing of this book being one of them. Lastly, the team at Quarry Books are also deserving of a huge Thank You for understanding my need for flexibility to still operate both my soap and beekeeping businesses while writing this book.

ABOUT THE AUTHOR

PETRA AHNERT is the owner and creative force behind Beehive Alchemy. She has been making soaps and body-care products for close to two decades, although the use of honey, beeswax and other bee-related products only came into the picture after she and her boyfriend discovered beekeeping in 2004 and a new business was born. She now sells her wares online (www.beehivealchemy.com) and at regional craft shows and farmers' markets. She lives in Milwaukee, Wisconsin, and keeps bees scattered throughout southwest Wisconsin.

INDEX